MW00713956

GENETICS

Nature's Blueprints

These and other books are included in the
Encyclopedia of Discovery and Invention series:

Airplanes: The Lure of Flight
Atoms: Building Blocks of Matter
Clocks: Chronicling Time
Computers: Mechanical Minds
Genetics: Nature's Blueprints
Germs: Mysterious Microorganisms
Gravity: The Universal Force
Lasers: Humanity's Magic Light
Movies: The World on Film
Photography: Preserving the Past
Plate Tectonics: Earth's Shifting Crust
Printing Press: Ideas into Type
Radar: The Silent Detector
Railroads: Bridging the Continents
Telescopes: Searching the Heavens
Television: Electronic Pictures

GENETICS

Nature's Blueprints

by LYNN BYCZYNSKI

The ENCYCLOPEDIA of
D·I·S·C·O·V·E·R·Y
and INVENTION

P.O. Box 289011 SAN DIEGO, CA 92198-9011

Library of Congress Cataloging-in-Publication Data

Byczynski, Lynn, 1954-
   Genetics : nature's blueprints / by Lynn Byczynski.
      p. cm. — (The Encyclopedia of discovery and invention)
   Includes bibliographical references and index.
   Summary: Discusses the history and development of genetic research, the discov-
ery of the gene, the relationship between genes and disease, genetic engineer-
ing, the ethics of this kind of research, forensic genetics, and the future.
   ISBN 1-56006-213-4
   1. Genetics—Juvenile literature.   [1. Genetics.]   I. Title.
II. Series.
QH437.5.B93   1991
575.1—dc20   91-15568

Printed in the USA

Contents

■■

Foreword   7

Introduction   10

CHAPTER 1 ■ Selecting the Seeds of a Better Crop   12
Early understanding of traits;
Learning how plants reproduce;
Cross-pollination and plant breeding;
The puzzles of human heredity.

CHAPTER 2 ■ Discovering the Basic Principles of Genetics   20
Charles Darwin's theory of evolution;
Gregor Mendel's experiments in heredity;
The results of breeding pea plants;
Obstacles to understanding.

CHAPTER 3 ■ Rediscovering Mendel's Lost Theories of Genetics   28
Using microscopes to study cells;
Rediscovering the theories of Mendel;
William Bateson spreads Mendel's ideas;
Searching for the key to heredity.

CHAPTER 4 ■ The Search for the Gene   39
Pneumonia's link to genetics;
DNA and heredity;
Discovering the structure of DNA;
The function of DNA.

CHAPTER 5 ■ Genetics and Human Health   49
The regularity of genetic mistakes;
Tay-Sachs disease and sickle-cell anemia;
Blood tests and prenatal care;
The link of cancer to oncogenes.

CHAPTER 6 ■ Genetic Engineering   57
    Altering genes that cause disease;
    The manipulation of DNA;
    Cloning rocks the scientific world;
    Using genes to make more food.

CHAPTER 7 ■ The Ethics of Genetics   68
    Bypassing life's basic rules;
    Different uses of genetic engineering;
    Sterilization in the U.S. and Germany;
    Genetic testing and the unborn child.

CHAPTER 8 ■ Fighting Crime with Genetics   76
    Using DNA to identify suspects;
    The development of DNA fingerprinting;
    Testing procedures raise questions;
    Challenging DNA as evidence.

CHAPTER 9 ■ The Future of Genetics   82
    Understanding the Human Genome Project;
    Innovative experiments with AIDS;
    Genetic advances in agriculture;
    The use of genetics in industry.

Glossary   87
For Further Reading   89
Works Consulted   90
Index   91
About the Author   95
Picture Credits   96

Foreword

The belief in progress has been one of the dominant forces in Western Civilization from the Scientific Revolution of the seventeenth century to the present. Embodied in the idea of progress is the conviction that each generation will be better off than the one that preceded it. Eventually, all peoples will benefit from and share in this better world. R.R. Palmer, in his *History of the Modern World,* calls this belief in progress "a kind of nonreligious faith that the conditions of human life" will continually improve as time goes on.

For over a thousand years prior to the seventeenth century, science had progressed little. Inquiry was largely discouraged, and experimentation, almost nonexistent. As a result, science became regressive and discovery was ignored. Benjamin Farrington, a historian of science, characterized it this way: "Science had failed to become a real force in the life of society. Instead there had arisen a conception of science as a cycle of liberal studies for a privileged minority. Science ceased to be a means of transforming the conditions of life." In short, had this intellectual climate continued, humanity's future would have been little more than a clone of its past.

Fortunately, these circumstances were not destined to last. By the seventeenth and eighteenth centuries, Western society was undergoing radical and favorable changes. And the changes that occurred gave rise to the notion that progress was a real force urging civilization forward. Surpluses of consumer goods were replacing substandard living conditions in most of Western Europe. Rigid class systems were giving way to social mobility. In nations like France and the United States, the lofty principles of democracy and popular sovereignty were being painted in broad, gilded strokes over the fading canvasses of monarchy and despotism.

But more significant than these social, economic, and political changes, the new age witnessed a rebirth of science. Centuries of scientific stagnation began crumbling before a spirit of scientific inquiry that spawned undreamed of technological advances. And it was the discoveries and inventions of scores of men and women that fueled these new technologies, dramatically increasing the ability of humankind to control nature—and, many believed, eventually to guide it.

It is a truism of science and technology that the results derived from observation and experimentation are not finalities. They are part of a process. Each discovery is but one piece in a continuum bridging past and present and heralding an extraordinary future. The heroic age of the Scientific Revolution was simply a start. It laid a foundation upon which succeeding generations of imaginative thinkers could build.

It kindled the belief that progress is possible as long as there were gifted men and women who would respond to society's needs. When Antonie van Leeuwenhoek observed *Animalcules* (little animals) through his high-powered microscope in 1683, the discovery did not end there. Others followed who would call these "little animals" bacteria and, in time, recognize their role in the process of health and disease. Robert Koch, a German bacteriologist and winner of the Nobel Prize in Physiology and Medicine, was one of these men. Koch firmly established that bacteria are responsible for causing infectious diseases. He identified, among others, the causative organisms of anthrax and tuberculosis. Alexander Fleming, another Nobel Laureate, progressed still further in the quest to understand and control bacteria. In 1928, Fleming discovered penicillin, the antibiotic wonder drug. Penicillin, and the generations of antibiotics that succeeded it, have done more to prevent premature death than any other discovery in the history of humankind. And as civilization hastens toward the twenty-first century, most agree that the conquest of van Leeuwenhoek's "little animals" will continue.

The *Encyclopedia of Discovery and Invention* examines those discoveries and inventions that have had a sweeping impact on life and thought in the modern world. Each book explores the ideas that led to the invention or discovery, and, more importantly, how the world changed and continues to change because of it. The series also highlights the people behind the achievements—the unique men and women whose singular genius and rich imagination have altered the lives of everyone. Enhanced by photographs and clearly explained technical drawings, these books are comprehensive examinations of the building blocks of human progress.

GENETICS

Nature's Blueprints

GENETICS

Introduction

When scientists first began studying genetics in the early 1900s, they knew they were pursuing a secret as old as humanity. People have always recognized that children have characteristics that are similar to those of their parents. And although people inherit certain features, variations and differences do appear. But throughout most of history, people could only wonder how heredity, the passing of features from one generation to the next, and variation occurred. No one understood the mechanisms that allow traits, or distinguishing features, to be inherited, or passed on.

The study of heredity, however, has developed rapidly. Scientists have learned that the secrets of heredity and variation are contained in microscopic structures called genes. They now know that genes are made up of a substance located in the cells. This substance, called DNA, contains the information that specifies all the characteristics that are passed from parents to offspring. In all forms of life, DNA is composed of the very same chemical ingredients.

This universal nature of genes has given researchers a common foundation

TIMELINE: GENETICS

1 ■ 300 B.C.
Aristotle proposes that genetic information is carried in the blood.

2 ■ 1676
Nehemiah Grew suggests that plants reproduce sexually.

3 ■ 1694
Rudolf Jakob Camerarius proves that plants reproduce sexually.

4 ■ 1859
Charles Darwin publishes his theory of natural selection.

5 ■ 1865
Gregor Mendel announces his discovery that inherited traits are carried on factors that divide into separate sex cells and recombine in later generations.

6 ■ 1884
Mendel dies without convincing other scientists he has discovered the mechanism of heredity.

7 ■ 1900
Mendel is rediscovered.

8 ■ 1905
Sex chromosomes are discovered.

in their quest to understand life. A knowledge of genetics is crucial in many areas of science. For example, the evolution of species cannot be explained without genetics. Diseases cannot be studied or treated without a consideration of genetics. The breeding of animals and plants relies heavily on genetics. Human behavior may even be largely influenced by genetics.

Despite the immense gains in knowledge during the last century, genetics is an infant science. Scientists are still mystified by many aspects of heredity. They know that thousands of diseases are inherited, but not which genes are responsible for all of them. Even when scientists identify defective genes, they do not know how to fix them. They know how genes work, but not what turns the genes on and off.

An army of scientists is working on these and other puzzles, and they are finding new answers daily. From food to medicine, increasing knowledge of genetics continues to benefit society. Because of the study of genetics, life today is far different than it was a century ago. Rapid advancements in the field now assure that life will be far different a century in the future.

7 8 9 10 11 12 13 14 15 16 17

9 ■ 1915
Thomas Hunt Morgan proves that genes are on chromosomes.

10 ■ 1943
Oswald Avery and colleagues discover that genetic information is carried in DNA.

11 ■ 1953
James Watson and Francis Crick discover the structure of DNA.

12 ■ 1961
The "genetic code" by which DNA makes proteins is revealed.

13 ■ 1970
Hamilton Smith discovers restriction enzymes, chemicals that cut DNA apart.

14 ■ 1972
Stanley Cohen and Herbert Boyer transplant a gene from a toad into a bacterium.

15 ■ 1980
Genetically engineered insulin is made available to diabetics.

16 ■ 1984
Harvard University scientists insert a human cancer gene into a mouse.

17 ■ 1990
Gene therapy in humans begins.

Selecting the Seeds of a Better Crop

The first application of genetics occurred ten thousand years ago, when primitive peoples realized they could plant some of the seeds they normally collected in the wild and grow whole fields of food. This process, called the domestication of plants, meant that societies of nomads, or wanderers, who had traveled constantly in search of food were able to settle down beside their fields and build villages.

Archaeologists think plants were domesticated first in the Fertile Crescent between the Tigris and Euphrates rivers in what is now Iraq. Crops grew abundantly in the rich soil of the Fertile Crescent, and soon the first great civilizations of Mesopotamia developed there.

The earliest farmers recognized that a plant's traits—its size, shape, color, and flavor, for example—were passed on in its seeds. They did not understand

This ancient Egyptian wall carving depicts farmers at work harvesting grain and tending cattle.

Early farmers, shown at work in this illustration, recognized that a plant's characteristics were passed on in its seeds. So they saved and re-planted the seeds of stronger plants and threw out the seeds of weaker ones.

how these traits were transmitted from one generation to the next; they simply knew by observation that this was occurring. They also knew by observation that many variations would appear in their crops. Some individual plants would be weak and small and fall prey to insects. Other plants would be bigger, more productive, tastier, or more insect-resistant. Because they knew that traits were carried in the seeds, the farmers would throw out the seeds of the weak, small plants. But they would carefully collect and save the seeds of the plants they found useful. The next year, when they planted these good seeds, the field would contain more useful plants. In

this way, unfavorable variations died out, and the favorable variations became more common. This process of planting seeds from the best plants is called selection. Over many years, crops were greatly improved through selection. Plants became bigger, hardier, and tastier, and they produced more food.

As crops improved, farmers were able to feed more people, and civilizations flourished. Better strains of wheat fed the blossoming nations of the Middle East and southern Europe. Improvements in rice supported the growth of Asian nations, and improvements in corn helped build South American nations. In addition to these three staple

By selecting seeds, farmers could grow bigger, hardier, and tastier crops. Improved corn, for example, helped build South American nations.

(Top) Thousands of years of genetic selection have produced these different corn strains. (Bottom) Italian physician Marcello Malpighi pioneered the scientific study of plant reproduction.

crops, many other wild plants were domesticated and improved through selection to provide a constant source of food. By about five thousand years ago, virtually every food crop eaten today was being cultivated somewhere in the world.

Plant Reproduction

Even with the great advances in food production, people still understood almost nothing about how heredity and variation worked. They did not even know how plants reproduced. And that information would not be discovered for thousands of years. But in 1672, an Italian physician, Marcello Malpighi, studied flowering plants under a microscope and published detailed observations of the anatomy and physiology, or structure and function, of plants. He noted that some flowers had small, threadlike organs called stamens, which produced a powdery substance called pollen. He found other flowers had a tubular organ called a pistil, which held the compartment where the fruit or vegetable was produced.

Four years later, a British physician, Nehemiah Grew, proposed that the stamen was a male organ and the pollen was the male sex cells. The pistils, he said, must be female organs containing the female sex cell—the ovum, or egg. Grew noted that most types of plants have flowers that contain both the stamen and the pistil, so they can fertilize themselves and then reproduce. In other types of plants, however, the male and female parts were located in separate flowers or even on separate plants, so the pollen from the stamen would have to be carried somehow to the pistil before reproduction could occur.

PLANT REPRODUCTION

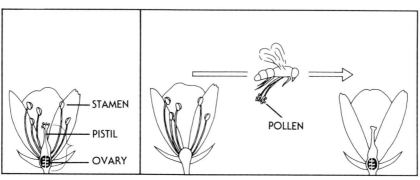

SELF-POLLINATION   CROSS-POLLINATION

STAMEN

PISTIL

OVARY

POLLEN

Most plants reproduce sexually, and their male and female reproductive parts are located in the flowers. The female reproductive part is called a pistil. The male reproductive part is called a stamen. The anthers, located at the tip of the stamen, produce pollen, a sticky powder that contains the sperm cells. A grain of pollen that falls onto the stigma, or top of the pistil, releases sperm cells. If a sperm cell reaches the ovary inside the pistil, it may fertilize, or unite with, the egg cell in the ovary to produce a seed for a new plant. This is called pollination. If the pollen that fertilizes a flower comes from the same flower or plant, it is called self-pollination.

The most common form of pollination, however, is cross-pollination. In this form of reproduction, the pollen from one plant must be carried to the flower of another plant by the wind, by hummingbirds, or by bees, butterflies, and other insects. Hummingbirds and insects are attracted to these flowers because they produce the sweet nectar that these creatures eat. As a hummingbird or insect enters a male flower to get the nectar, some pollen sticks to it. Then, when it flies into a female flower for more nectar, some of the pollen it is carrying falls onto the pistil.

Some plants can either cross-pollinate or self-pollinate. Others, such as willows and poplars, can only cross-pollinate because they produce separate male flowers, which contain no pistil, and female flowers, which contain no stamens.

Many people, particularly religious officials, rejected Grew's ideas about plant reproduction. In the fifteenth century, people did not talk about sex, and the thought that plants had a sexual function was considered shocking and immoral. Despite the disapproval that Grew received, other scientists decided to study the reproduction of plants. In 1694, the German scientist Rudolf Jakob Camerarius published the first evidence of plant sexual reproduction. Camerarius removed the stamens from male flowers of the castor-oil plant to see if the plant could reproduce without them. When the female flowers failed to produce seeds, Camerarius concluded that flowers needed both stamens and pistils to reproduce. He also concluded that flowers containing both stamens and pistils pollinated, or fertilized, themselves. Other flowers had to have the pollen carried from the stamen to the pistil either by the wind or by insects. This transfer of pollen from one flower to another came to be known as cross-pollination. Soon, other researchers found that cross-pollination could also occur between different types of plants. This was an exciting development because it meant that characteristics of two different types of plants could be combined to make a new plant.

Breeding Plants

With this knowledge, the business of plant breeding was born. Breeders studied the reproductive cycle in plants and began experimenting with cross-

By the time Christopher Columbus arrived in the Americas, Mexicans were growing many varieties of tomatoes.

pollination. They soon learned that they could pick up pollen from one plant on a tiny paintbrush and then dust the pollen on the pistil of another plant. With plants that contained both stamens and pistils, they would pluck out the stamen before it had a chance to produce pollen. Then, they would dust pollen from another plant on the pistil of the first plant. Finally, they would tie a small bag around the pollinated flower to make sure that no other pollen could be brought into the flower by wind or insects.

The process of cross-pollinating different types of plants is known as hybridization. The offspring of cross-pollinated plants are called hybrids. Using hybridization, plant breeders were able to speed up the process of selection that had improved crops over hundreds of years. Once they knew how to combine traits they liked, they could create new and improved varieties of plants within a few years. They could make plants that were stronger, tastier, and more productive. For example, the first tomatoes were tiny, bitter fruits on scraggly vines growing in South America. Over many generations, tomatoes became bigger and sweeter because people selected and planted only the best seeds. Sometimes, a much different plant, one with bigger fruit or fruit of a different color, for example, would appear in the field. Although farmers did not know why this occurred, they would save the seeds of plants with desirable variations. In this way, different varieties of tomatoes developed. By the time Christopher Columbus came to the Americas in 1492, Mexicans were growing many varieties of tomatoes with different shapes and a range of colors that included orange, yellow, white, and red.

In the 1800s, tomatoes became popular in North America, and plant breeders began to apply the knowledge they had gained about cross-pollinating as a way of combining desirable characteristics. Instead of waiting for a better tomato to appear spontaneously, which might never happen, breeders decided to combine the characteristics they wanted in a tomato. By the middle of the century, dozens of new varieties had been developed that combined the best traits of the old varieties.

As plant breeders cross-pollinated different combinations of plants, they learned that some plants could not be cross-pollinated. They just would not produce offspring. These discoveries contributed to the development of the idea of what a species is. For hundreds of years, people had tried to classify animals and plants into distinct groups

Early people noticed that different seeds from the same type of plant sometimes produced different-sized plants. But what, they wondered, caused similar variety in humans?

Originally small and bitter, the tomato has been hybridized into a luscious fruit that now comes in a variety of shapes, sizes, and colors.

based on their similarities. Eventually, in the 1700s, scientists settled on a system that defined a species as a group of organisms that can breed together to produce fertile offspring that have the same characteristics as the parents. This definition of species remains valid today.

Pondering Human Variety

During the thousands of years that people were putting selection and breeding to work in their fields, they also were thinking about the puzzle of human heredity. They knew that sometimes traits were passed from one generation to the next, as when a son looked just like his father. They knew that children would sometimes have a mixture of their parents' features—a mother's curly hair and a father's cleft chin, for example. They knew that parents could have several children who did not look at all like one another. Sometimes, they saw a trait skip a generation and reappear in

(Right) Physical traits from parents are often passed on to their children, resulting in families whose members look noticeably alike, as in this picture of a mother surrounded by her daughters.

(Right) Seventeenth-century British physician William Harvey disproved Aristotle's (above) long-accepted particulate theory of heredity.

grandchildren. And sometimes, they saw children with traits unlike anything ever seen before in the family.

These observations caused a great deal of speculation among philosophers and scholars throughout history. About 300 B.C., the Greek philosopher Aristotle proposed that the blood of humans and animals carried particles that determined the characteristics of their offspring. He thought that the particles blended in the offspring, like cream and coffee, creating traits that were a combination of the parents' traits. A tall father and short mother would produce a child of medium height, for example. Aristotle's beliefs about heredity were called the "particulate theory," and this theory dominated scientific thinking for hundreds of years.

Then, during the mid-1600s, the British physician William Harvey studied the development of an egg into a chick and disproved the idea that the parents' blood mixed inside the egg. But Harvey

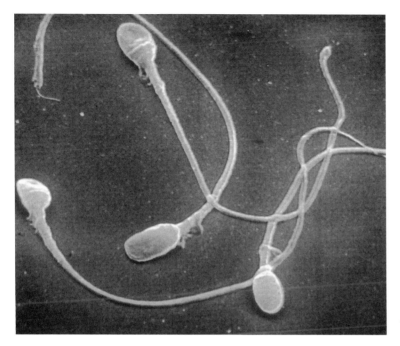

Some eighteenth-century scientists claimed they could see tiny babies curled up inside human sperm cells through their crude microscopes. Although these observations were incorrect, the scientists did disprove William Harvey's century-old hypothesis that the father had no role in reproduction.

sent science down another dead-end path when he claimed that the father had no role in creating offspring. He said that all the material necessary for a new life was contained in the mother's egg. Harvey's theory was contradicted a century later by yet another incorrect theory about inheritance. Several scientists studying sperm, the male reproductive cell in humans and animals, under the crude microscopes of the day claimed they could see tiny babies curled up in the sperm cells.

All of these wrong theories about heredity were based entirely on observation. Scientists did not test their observations with experiments, as scientists do today. Those who wrote about inheritance through the 1700s were really just guessing. Useless as their ideas turned out to be, their work at least showed that people were keenly interested in heredity. It was not until the middle of the 1800s, when a more careful approach to science developed, that some of the facts of heredity were revealed.

Discovering the Basic Principles of Genetics

In 1859, the British scientist Charles Darwin shook up the scientific world with his book *The Origin of Species,* in which he suggested that new species develop over time in a process called evolution. Darwin's theory was extremely controversial because most people of his day believed that God had made all species on the third day of creation and that living creatures had not changed since then.

Darwin's Theory

To support his theory of evolution, Darwin pointed out that individuals within a species are not exactly the same. Many individuals show variations in characteristics such as body size and hardiness. Sometimes, variations can help an individual survive longer and produce more offspring than other members of

the species. Such individuals would pass down to their offspring the variation that helped them survive. Over time, the individuals with this favorable variation

British scientist Charles Darwin (above) theorized that nature selected the best variations in a species just as farmers had done with plant seeds. Plants, with all their different varieties, provided a starting place to study this theory further.

Darwin believed that the development of the peacock's gorgeous tail feathers was a prime example of natural selection.

would become more numerous than those without it. Eventually, an entirely new species with the variation would result. Darwin called his theory "natural selection" because he thought that nature selected the best variations in a species, just as early farmers had selected the plants that best suited their needs. Other scientists called Darwin's theory "survival of the fittest," and that phrase is still used today to describe how species evolved.

Darwin cited the peacock as an example of natural selection, or survival of the fittest. Male peacocks attract mates by displaying their colored tail feathers. A male peacock with especially colorful feathers would be so attractive to females that he would attract more mates and produce more offspring. In this way, Darwin said, the especially colorful tail feathers would be passed on to more peacocks. After many generations, most male peacocks would have these colorful tail feathers.

Darwin's theory about evolution rested on the common knowledge that parents pass traits on to their children. But during Darwin's time, no one knew anything about how this process took place. Once Darwin focused on heredity's role in evolution, many scientists turned their attention to the mystery of how heredity works.

Mendel's Pea Plants

Among the many scientists trying to solve the mystery of heredity was a monk named Gregor Mendel. Mendel, born in 1822 to a peasant family, had few scientific credentials. His chief job at his monastery in Brunn, Austria, was to tend the vegetable and flower gardens. But he was an experienced plant breeder, and he suspected that garden plants might give him an understanding of the laws of heredity. Mendel and other plant breeders of his time knew several things about heredity in plants. First, they knew that a single species could have many variations in size, flower color, seed shape, taste, productivity, and other characteristics. They also knew that if they mated plants with different traits, such as tall plants and short plants, they might get all tall plants in one generation, then a mixture of tall and short plants in the next generation. What plant breeders,

Austrian monk Gregor Mendel's nineteenth-century breeding experiments with pea plants laid the foundation for modern genetics.

er color, height, and seed shape, among other traits. Mendel also knew from previous experience that pea plants are easy to breed and that their breeding can easily be controlled. The tiny, tightly closed pea flowers make it unlikely that insects or wind will bring pollen from foreign plants. Mendel was familiar with the hybridization procedure of plucking out the stamen in one plant and transferring pollen from another plant on a tiny brush. Using that procedure, he successfully mated plants with different traits.

Mendel decided he would study seven characteristics of the pea plant: flower color, seed shape, seed color, pod shape, pod color, height, and location of flowers on the stem. He mated plants with purple flowers and white flowers, wrinkled seeds and smooth seeds, yellow seeds and green seeds, puffy pods and flat pods, and green pods and yellow pods. He also mated tall plants with short plants, and he crossed plants whose flowers grew on the side of the stem with plants that had flowers at the end of the stem.

including Mendel, did not know was why or how that happened.

In the spring of 1856, Mendel decided he would learn how plants passed on their traits to their offspring. To do this, he knew he would have to watch plants reproduce for several generations. Only in this way, would he be able to see how often the traits of a parent plant would appear in subsequent generations. For his experiments, Mendel chose pea plants. He chose peas because they grew well at the monastery and matured quickly. He knew that peas showed a great deal of diversity in flow-

The tightly closed flower of the pea plant prevents natural cross-pollination, making it a good candidate for plant-breeding experiments.

Mendel worked with pea plants for eight years to understand the workings of heredity.

Mendel did these crossbreeding experiments by dusting pollen from one plant onto another. He dusted pollen from a purple-flowered plant onto a white-flowered plant, for example. He let these plants grow to maturity, then collected the seeds. The next year, he planted the seeds and raised a new crop of peas. All of these plants, called the first generation, had purple flowers. This time, instead of brushing the flowers with pollen from other flowers, Mendel let all the purple-flowered plants pollinate themselves naturally. When they matured, he saved the seeds they produced and planted them the next year. This batch of seeds produced a crop of pea plants known as the second generation. Most of the plants in the second generation had purple flowers, like their parents. But some had white flowers, like the grandparent plants.

Mendel repeated this experiment with each of the other six pairs of characteristics. Although the characteristics were different, the results were always the same. Just as a cross between purple and white flowers produced all purple flowers in the first generation and a mixture of purple and white flowers in the second generation, Mendel found that tall and short plants produced tall plants in the first generation and a mixture of tall and short plants in the second generation.

Counting the Results

Other plant breeders before Mendel had noticed this same pattern. But Mendel did something that set him apart from all the others working on the mystery of inheritance: He counted the plants in the second generation that exhibited the characteristics he was studying. He counted how many had purple flowers, how many had white flowers; how many had round seeds, how many had wrinkled seeds; and how many were tall, how many were short. This by itself was quite unusual. Biologists at that time did not keep careful count of the results of their experiments. Because Mendel did count his results, he was able to find a pattern in the way traits reappeared. He then applied his knowledge of mathematics to determine what caused the traits to disappear in the first generation and reappear in the second.

When he counted his results, Mendel got the same number in each of his experiments. For every three plants with purple flowers, he got one plant with white flowers. For every three plants with smooth seeds, he got one plant with wrinkled seeds. The numerical relationship between two or more things, such as purple flowers and white flowers or smooth seeds and wrinkled seeds, is called a ratio. In each of Mendel's experiments, the traits occurred in a 3:1 ratio in the second generation. But what did this 3:1 ratio mean?

Mendel realized that the constant 3:1 ratio provided a mathematical clue to the way characteristics are passed from parents to offspring. He figured out that each parent plant must contain some element that determines the characteristics of its offspring. This "element" later came to be known as a gene. But, according to Mendel, each parent can have two versions of the same gene, and each version produces slightly different results in the offspring. The characteristics of the offspring depend on which version of

Mendel carefully controlled and documented his experiments with pea plants.

the gene it inherits. These different versions of the same gene were later named alleles. For example, the gene that controls flower color in pea plants has two alleles, one for purple flowers and one for white flowers. And each plant contains a pair of alleles in its cells. Therefore, a plant can have two alleles for purple flowers, two alleles for white flowers, or one allele for each. When the parent plant forms sex cells—either pollen or eggs—each sex cell receives only one of its parents' two alleles. When the male and female sex cells join in the fertilization process, the new plant that is created will have one allele for flower color from the male parent and one allele for flower color from the female parent.

Because one trait—purple flowers, for instance—showed up more often than the other, Mendel also realized that one of the alleles must be dominant, or stronger, and the other must be recessive, or weaker. When purple-flowered plants were cross-pollinated with white-flowered plants, the purple was dominant and the white was recessive because all the offspring had purple flowers. But that recessive white allele was still present in the first generation of purple-flowered plants. And when two parent plants passed their recessive white alleles to offspring in the second generation, the allele for white flowers was not overpowered by the dominant purple allele. As a result, white flowers appeared in the offspring.

These conclusions were important because they revealed a system of inheritance that plant breeders could use as they tried to develop new varieties of plants. According to Mendel's theory, plants would always exhibit the dominant trait in the first generation, and a 3:1 mixture of dominant and recessive traits in the second generation.

RESULTS OF MENDEL'S PEA PLANT EXPERIMENT

Because genes were invisible to Mendel and other people of his time, Mendel devised a diagram to help others visualize the action of the alleles. In his diagram, he assigned letters to each characteristic and used a capital letter to designate a dominant trait and a lowercase letter to designate a recessive trait. A capital *A* identified the dominant allele for purple flowers in pea plants, for example. A lowercase *a* identified the recessive allele for white flowers. If a parent plant had purple flowers, it would be identified as "AA." If the parent plant had white flowers, it would be identified as "aa." Mendel thought that each sex cell received only one allele from the parent, so the sex cells would be designated as "A" or "a." Here is how he showed the results of crossing plants with purple flowers (AA) with white flowers (aa):

AA x aa

	A	A
a	Aa	Aa
a	Aa	Aa

Aa x aa

	A	a
a	Aa	aa
a	Aa	aa

Revealing the Findings

Mendel was unable to see the genes that determine characteristics because they are so tiny and hidden within individual cells. Mendel could really only guess at the workings of whatever was responsible for controlling inherited characteristics.

But after eight years of studying pea plants, Mendel felt confident that he finally understood how heredity works, even if he did not know anything about the genes that control the process. He was ready to share his discovery with the world. The first step was to present his findings to other scientists for review.

Hawkweed proved a poor substitute for pea plants in Mendel's later experiments. The hawkweed is capable of reproducing without mating which means sex cells are not involved.

Mendel thought his experiments had turned up important results that would create both interest and skepticism in the scientific community. He fully expected his fellow scientists to vigorously question, debate, and test his findings. In February and March 1865, Mendel presented a lecture on his results to the local Natural Science Society in Brunn. He informed its members of his three conclusions. First, he said, factors that control hereditary characteristics exist in the cells and split apart when sex cells are formed so that each sex cell has only half of the information. Second, one characteristic in a pair is dominant and one is recessive. And third, the factors that control the inheritance of characteristics are separate and can combine in different ways. In the second generation, this combination process allows the recessive trait to reappear.

Although Mendel's conclusions were certainly new and important, the scientists at the meeting did not ask one question. No one understood the significance of what Mendel was saying. No one realized that what he had done was observe the visible results of a plant-breeding experiment, then figure out the invisible workings of the cells that were involved in reproduction. They did not realize that Mendel was offering them the first real clue to life's most basic process. They did not even see that Mendel's rules could help plant breeders in their work by predicting in advance the results of cross-pollination. Mendel had explained the hidden process by which traits are passed on, but none of his fellow scientists understood this.

Even when he published his results in a scientific journal, Mendel received no response. So, he sent a copy of his paper to Carl Nageli, the most famous botanist in Europe, in hopes of getting Nageli to confirm his results. Nageli apparently did not grasp the significance of the work either, and he wrote back to Mendel that he viewed it "with mistrustful caution."

For the next seven years, Mendel and Nageli corresponded, exchanging plants and reporting results of further experiments. This friendship turned out to be bad luck for Mendel. In 1867, he had to give up his work with peas because the pea beetle had invaded the monastery gardens and eaten all his plants. At Nageli's recommendation, Mendel tried to repeat his experiments on a plant known as hawkweed. Using this plant, Mendel never got the kind of results he had obtained with peas, and he even began to doubt his theories about inheritance. Many years later, scientists discovered that hawkweed is one of the few plants that often produces offspring without mating, which means sex cells are not involved. Mendel's theory that characteristics are passed to offspring through the sex cells would not apply to that kind of plant.

Gathering Dust

By this time, Mendel had begun to do fewer breeding experiments. As he wrote to Nageli, "I am no longer very fit for botanical field trips." It was the monk's way of saying he had gotten too fat to climb mountains in search of plants.

In 1868, Mendel was chosen to be the abbot of his monastery. His new duties took him away from the monastery for long periods of time, and in 1873, he gave up his work with plants altogether. His notes and research papers gathered dust in the monastery's library, and on January 6, 1884, Mendel died of a kidney disease at the age of sixty-two.

It took sixteen years before anyone else really understood that Gregor Mendel had discovered the basic principles of genetics.

Rediscovering Mendel's Lost Theories of Genetics

In the years after Mendel's death, the most exciting scientific discoveries were occurring under the microscope. As it turned out, this research was exactly what was needed to make the world ready for a rediscovery of Mendel's theories.

A primitive sort of microscope had been invented in the 1590s, but the device was considered a toy, not a scientific tool. In 1665, though, the British scientist Robert Hooke took the microscope more seriously. He began studying all kinds of materials through it and writing reports of what he saw. A thin slice of cork, he wrote, was made up of

Microscopes enabled scientists to study cells and cell reproduction. This work in turn led to the rediscovery of Mendel's theories. At right, are plant cell nuclei visible under modern microscopes.

"a great many little boxes" that he called cells. The name stuck.

For the next two hundred years, people studied cells under microscopes. They argued about what they saw but eventually agreed on several facts. First, they concluded that all living organisms are made up of cells and that all cells have a central structure called a nucleus. Second, they agreed that organisms grow when cells split to form two new cells, each with its own nucleus. Third, they realized that when sex cells, which are called gametes, join together to form a new cell, the new cell begins to divide immediately. The researchers recognized this as the birth of a new organism.

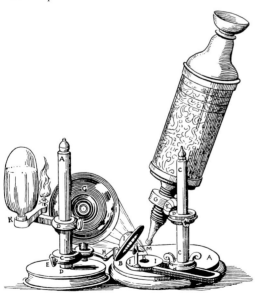

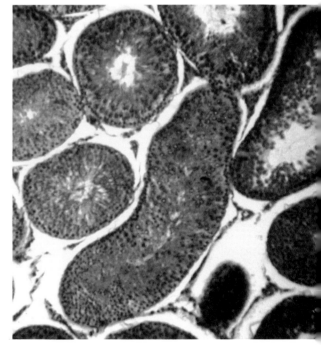

MITOSIS

Mitosis is the process of cell division that makes all living things grow. In this process, a single cell divides into two cells that have exactly the same number of chromosomes as the original cell. The two new cells then repeat the process, each one producing two more new cells, and so on. Information contained in the chromosomes directs some cells to develop into muscle cells, others into blood cells and other specialized cells. The phases of mitosis are shown below.

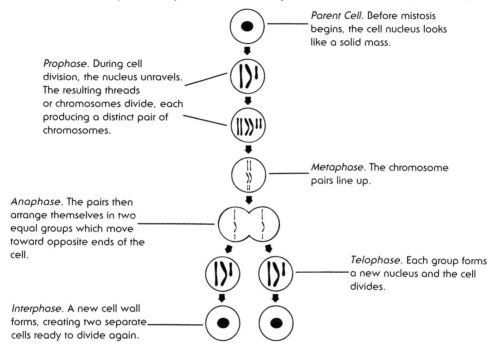

Parent Cell. Before mistosis begins, the cell nucleus looks like a solid mass.

Prophase. During cell division, the nucleus unravels. The resulting threads or chromosomes divide, each producing a distinct pair of chromosomes.

Metaphase. The chromosome pairs line up.

Anaphase. The pairs then arrange themselves in two equal groups which move toward opposite ends of the cell.

Telophase. Each group forms a new nucleus and the cell divides.

Interphase. A new cell wall forms, creating two separate cells ready to divide again.

Seeing Cells Divide

A microscope that provided a clearer, bigger image was developed in 1877, and cell research became even more popular. Scientists could actually see what was happening inside the cell as it divided. As they watched cells divide, they noticed a peculiar thing occurring. The nucleus of the cell seemed to fade away. Then, several coiled threads would appear, split in two, and move apart. Then, the threads faded and two new nuclei would appear just as the cell divided. Each new cell would have its own nucleus, which contained a full set of these mysterious threads.

Scientists named this cell-division process mitosis, from the Greek word for "thread." They named the threads chromosomes, from the Greek word for "color," because the threads could be seen only when stained with a dye.

As they studied dividing cells under a microscope, scientists came to realize that each type of organism has a constant number of chromosomes. All pea plants, for example, had fourteen chromosomes in their cells. A small

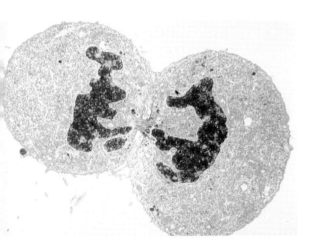
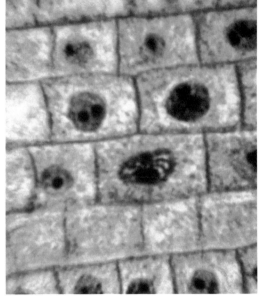

(Left) A cell splits during mitosis to form two new cells, each with its own nucleus. (Right) The tell-tale "threading" of the nuclei of some of these plant cells reveals that they are undergoing mitosis.

worm that was often studied under the microscope had only two chromosomes. Human cells had forty-six chromosomes. When a cell split during mitosis, the number of chromosomes would double and then divide evenly when the two new cells pulled apart. As a result, each new cell would have the same number of chromosomes as the original cell.

But then the scientists noticed something else. When cells divided to form sex cells, something different happened. The number of chromosomes did not double before the cell divided, but did split evenly, which meant that each sex cell got only half the number of chromosomes in the parent cell. That made sense to the researchers. They realized that if every male gamete

The forty-six chromosomes of a normal human female arranged for study purposes.

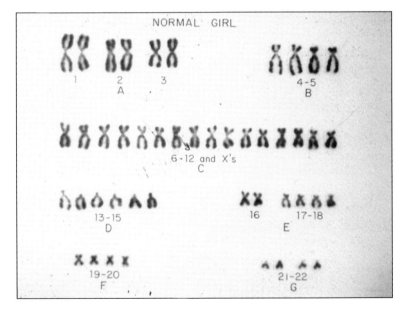

MEIOSIS: THE BIRTH OF AN ORGANISM

Sex cells, or gametes, divide differently than all other living cells. This process is called meiosis. Instead of producing two new cells that contain full sets of chromosomes, a sex cell produces two special cells. A gamete contains only half of an original set of chromosomes, and it cannot divide like other cells. A male gamete, or sperm cell, must unite with a female gamete, or egg cell, to form the first complete cell of a new offspring, called a zygote. Because the zygote receives half its chromosomes from each parent, it will inherit some genetic characteristics from its father and some from its mother.

When sex cells divide, each cell splits into two gametes that contain only half of the original set of chromosomes, as shown in the diagram.

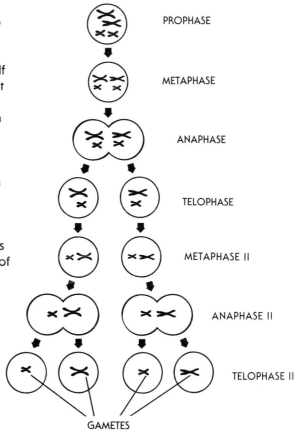

PROPHASE

METAPHASE

ANAPHASE

TELOPHASE

METAPHASE II

ANAPHASE II

TELOPHASE II

GAMETES

and every female gamete had a full number of chromosomes, the new cell produced would then have twice the number of chromosomes. And that number would double every generation. Yet scientists knew that was not occurring because they had studied generations of organisms, and each generation had the same number of chromosomes. This process of cell division that created gametes was named meiosis, from the Greek word meaning "to diminish."

While some scientists continued to study the cell, several others were still working on the mystery of why members of the same species could have many differences in appearance. Several researchers were even looking at the same kind of plant Mendel had originally studied. Although these various scientists had never heard of Mendel or his pea plant experiments, they had noticed that some pea plants had white flowers and some had purple flowers, and like Mendel, they wondered why.

Botanist Hugo De Vries reached the same conclusions about heredity as Mendel thirty-five years earlier. De Vries was disappointed to learn that he was not the first to discover the mechanism of heredity.

Keeping Track of Traits

One of these scientists, the Dutch botanist Hugo De Vries, was breeding a type of plant called silene in 1893. He cross-pollinated a variety that had smooth leaves with a variety that had hairy leaves and found that all the offspring had hairy leaves. He let those plants pollinate naturally among themselves and found the second generation had mostly hairy leaves, but some had smooth leaves. He counted how many of each and got a ratio of three plants with hairy leaves to one plant with smooth leaves.

It was the same ratio Mendel found thirty years earlier in his pea plant experiments. And like Mendel, De Vries

De Vries conducted his experiments with flowering silene plants, just as Mendel had used pea plants.

realized the implications of the ratio. He figured out that there must be two different factors controlling the appearance of the silene plant leaves—one a factor for hairy leaves and one a factor for smooth leaves.

He studied the 3:1 ratio and did the same mathematical calculations that Mendel had done. His math told him that the two separate factors controlling leaf texture must separate into different sex cells. Then, when the sex cells joined, the factors would combine in different patterns in the offspring. He also thought that one factor must be dominant over the other because when he mated plants with hairy leaves and plants with smooth leaves, he got a first generation of plants with hairy leaves and a second generation with both hairy and smooth leaves. This told him the plant would exhibit the dominant factor—in this case, hairy leaves—unless it got a double dose of the recessive factor. If that happened, then the plant would take the appearance of the recessive factor and produce smooth leaves.

These results made De Vries think he had discovered the invisible mechanism by which traits are passed from one generation to the next. He thought he had solved a mystery that had puzzled people for thousands of years. And he hoped that he would be remembered as the scientist who explained this most basic process of life.

De Vries tested his ideas for seven years. Finally, he felt sure that inherited traits were determined by some kind of physical factors that separate in the sex cells, then recombine in the offspring. As he was writing the paper that would announce his discovery to the world, De Vries received a big envelope from a friend. This friend had found a copy of an old scientific paper that he thought might have some relevance to De Vries's work. Enclosed was a copy of Mendel's 1865 paper on his experiments with pea plants.

Rediscovering Mendel

As De Vries read Mendel's paper, his heart sank. He had never heard of Mendel, but he quickly realized that his own theories on inheritance were old news. A monk had arrived at the very same conclusions thirty-five years earlier. Nevertheless, De Vries finished his paper and included several references to Mendel's pea experiments. He thought that he would at least become famous as the rediscoverer of Mendel's theories.

De Vries soon learned he had to share even that honor. Two other scientists, Carl Correns in Germany and Erich von Tschermak in Austria, had also discovered Mendel's paper and had applied it to their own experiments. They had even done their work on pea plants.

In 1900, all three men separately published their papers. Finally, it appeared Mendel would receive the recognition he deserved.

On May 8 of that year, the British scientist William Bateson was catching up on some reading as he took a train to a scientific meeting. In one of his magazines, he read a summary of Mendel's paper in an article on De Vries's work. Bateson later told friends he had experienced one of the most dramatic moments of his life. Here was the answer he had been seeking. For years, Bateson had been studying inheritance but had never really figured out how the hereditary mechanism worked. Mendel's concise paper immediately cleared the uncertainty that had bothered Bateson. Mendel had shown him that the instructions for inherited characteristics must be contained in physical factors that separated into sex cells, then recombined in different ways when the sex cells joined to form a new life.

From that day on, Bateson worked at spreading Mendel's ideas through the world of science. He believed that if other scientists continued Mendel's work, they would soon learn exactly

De Vries hybridized silenes with hairy leaves and silenes with smooth leaves in his genetics experiments.

"X" AND "Y": THE MALE/FEMALE CHROMOSOMES

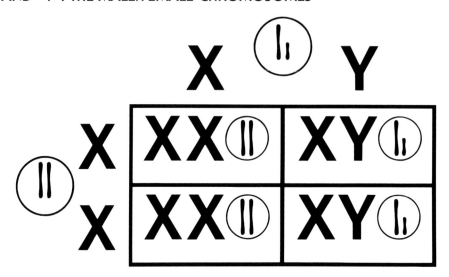

All of the cells of a female organism contain an identical pair of "X" chromosomes. In male organisms, however, an "X" chromosome is paired with a shorter "Y" chromosome. This "Y" chromosome carries the genes that cause the organism to be a male.

A male sperm cell, however, contains only one chromosome. This single chromosome may either be an "X" chromosome or a "Y" chromosome. The female egg cell, on the other hand, can contain only an "X" chromosome. As this diagram shows, if the egg combines with a sperm cell, it has an equal chance of gaining another "X" chromosome, making the offspring female ("XX"), or of receiving a "Y" chromosome, making the offspring male ("XY").

how these instructions are carried from parents to offspring. Like most scientists, Bateson was driven by a desire to understand the processes governing life. But he also knew that this knowledge could eventually lead to discoveries that would benefit society. It was Bateson who named the new science of heredity genetics and standardized the use of the word *gene* to describe the factors that were carried in the sex cells. He also coined the word *allele* to describe the dominant and recessive genes for the same characteristics.

Bateson also conducted his own experiments with poultry and found that some traits were inherited according to the same pattern Mendel had found in pea plants. His work proved that Mendel's theories applied to animals as well as plants. This was a first step toward understanding that the process of heredity is the same in all living things.

Understanding Chromosomes

As Bateson carried on his public relations campaign for Mendel's work, he caught the attention of some of the scientists who had been studying cell division. Mendel, who did his work before powerful microscopes were invented, could only guess that there were physical factors inside the cell that separated during meiosis. Those whose powerful new microscopes showed the dividing cells realized that Mendel's "factors" could be the chromosomes. These scientists joined in the search for the mechanism of inheritance.

By 1905, researchers had noticed that chromosomes were linked in pairs. In most pairs, each half was the same size and shape. But some cells had an odd pair of chromosomes, with one long chromosome and one tiny chromosome. Scientists called the long chromosome in the odd pair an X chromosome and the tiny chromosome the Y chromosome. Then, they realized that the XY combination occurred only in males. Females of the same species had two X chromosomes. The scientists realized that the sex of an organism was determined by this special pair of chromosomes.

This discovery was important because it provided the first solid evidence that chromosomes were somehow involved in the process of heredity. But this evidence was not enough. Scientists were inspired to study the chromosomes even more closely to see what else they could learn about genetics. And many scientists all over the world took up the challenge.

Among them was a group of science professors and students at Columbia University in New York City. These researchers wanted to prove that chromosomes were involved in the inheritance of characteristics. So far, the only connection that had been established between genes—these invisible factors that determine the inheritance of char-acteristics—and chromosomes was that the X and Y chromosomes determined sex. The Columbia scientists realized that if chromosomes did contain the genes, each chromosome must have many genes on it. That much seemed obvious to them. Humans have only forty-six chromosomes, but they have many thousands of characteristics that are inherited. Futhermore, the scientists theorized that if many genes are connected on a chromosome, the entire package of genes would be inherited together. This theory was supported by microscopic studies of meiosis, which showed that each chromosome passed as a whole into the gametes.

Testing the Theory

Those were the theories that guided the Columbia researchers. Proving those theories was a difficult matter that would occupy them from 1909 until the 1930s. This group of scientists decided to test their theories by studying the chromosomes of a tiny insect called *Drosophila melanogaster,* or the fruit fly. They chose fruit flies because the insects took up little space and produced a new

Drosophila melanogaster, *the common fruit fly, became the preferred species for genetic research because of its brief reproductive cycle.*

generation every two or three weeks. With such a short reproductive cycle, researchers could quickly see what characteristics were passed to the offspring.

The first evidence that chromosomes contain more than one gene came from the work of Thomas Hunt Morgan, one of the scientists at Columbia. Morgan found a fruit fly with white eyes, quite different from the usual red-eyed fruit fly. He mated the white-eyed fly with a red-eyed fly. All the offspring had red eyes, so Morgan assumed red eyes were dominant over white eyes. Then, in the same way

Columbia University geneticist Thomas Hunt Morgan proved that genes were carried on chromosomes.

Mendel had let his first generation of peas pollinate naturally, Morgan let the first generation of fruit flies mate. Among their offspring, he found three red-eyed flies for every one white-eyed fly. These results told Morgan that the allele for white eyes must be recessive, and the allele for red eyes dominant. It followed the same pattern of inheritance displayed by Mendel's pea plants and De Vries's silene plants.

As Morgan looked more closely at these white-eyed flies, however, he discovered they were all males. He thought for a long time about why this might have occurred. He knew that males have an X chromosome and a Y chromosome. And he knew that the allele for white eyes was recessive.

Suddenly, Morgan had an idea about the invisible workings of the chromosome. The gene for eye color, he concluded, must be carried on the X chromosome. Here is how he came to this realization: Female offspring have two X chromosomes, one from each parent. An X chromosome from a white-eyed parent would contain the recessive allele for white eyes, and an X chromosome from a normal red-eyed parent would contain the dominant allele for red eyes. Those two alleles would be contained in the XX female offspring, and the dominant red allele would overpower the recessive white allele. Male offspring, however, get one X chromosome from one parent and one Y chromosome from the other parent. If the X chromosome came from a white-eyed parent, it would carry the recessive white allele. The Y chromosome did not carry any gene for eye color, Morgan concluded, so the recessive white allele on the X chromosome would show up in the XY male offspring.

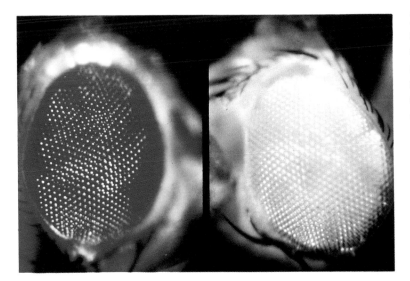

The normal eye color of the common fruit fly is red (left). The white eye on the right is a recessive trait carried only by a gene on an X chromosome.

Beads on a String

By showing that a genetic trait was linked to a specific chromosome, Morgan became the first to prove that genes are part of chromosomes. Later, Morgan proved that each chromosome contains many genes and that the genes are lined up on the chromosomes like beads on a string. These discoveries were important in the developing field of genetics because they helped scientists understand where the mechanism of heredity is located. Scientists began to focus their studies on the chromosomes.

While Morgan was working on the location of genes, one of his students, Hermann J. Muller, was tackling another perplexing question. Ever since Darwin had proposed that new species evolve through the natural selection of the most favorable variations, people had wondered what caused those variations to appear in the first place. Variations were certainly common. Farmers found odd plants in their fields all the time and either cultivated or destroyed the plants, depending on

whether the variation was useful to them. Animals, too, would sometimes produce offspring with an unusual characteristic that neither parent possessed. The white-eyed fruit flies in Morgan's laboratory were an example. These slight changes, called mutations, seemed to occur spontaneously.

But Muller did not want to wait around for a spontaneous mutation to occur so that the *Drosophila* researchers could test its pattern of inheritance. Muller wanted to know if he could cause mutations to occur. He tried exposing *Drosophila* to high temperatures, thinking that the heat might somehow disrupt the orderly process of inheritance. That did not happen, though, so he looked for some other way to assault the flies' genes. By this time, scientists already knew that living tissues could be harmed by too much exposure to X rays, the radiation used to take pictures of bones through the skin. Muller decided to bombard his fruit flies with X rays. The technique worked. The flies exposed to X rays produced offspring with variations that were not seen in the parents. In some

Hermann J. Muller discovered that environmental factors, such as radiation, could cause genetic changes, or mutations, and produce new traits.

cases, Muller could even see through the microscope that changes had occurred in the chromosomes of the flies exposed to X rays.

Muller's work proved that environmental factors can cause chromosomes to change, which, in turn, causes changes in inherited characteristics. The work of Muller and Morgan convinced most scientists that genes would be found on the chromosomes. But many more years would pass and thousands of experiments would be conducted before the structure of the gene would be revealed.

CHAPTER 4

The Search for the Gene

Many times in the history of science, important discoveries have been made by accident. A good scientist, though, is alert to the possibilities offered by the unexpected. That is how Frederick Griffith, a British scientist, made his contribution to genetics in 1928.

Griffith was trying to find a cure for pneumonia, a lung disease that in the first half of the century was often fatal. Research had shown that pneumonia was caused by a type of bacteria, which are simple, one-celled organisms. Bacteria can be helpful as well as harmful, but the type of bacteria Griffith was studying was definitely harmful. It was taken from people who had died of pneumonia.

Mysterious Changes

As he examined the bacteria under a microscope, Griffith noticed that most of them had a smooth protective coat. A few, however, were rough in appearance. Griffith wanted to know whether the two types of bacteria had different effects on animals, so he injected some mice with the smooth bacteria and other mice with the rough bacteria. The mice injected with the smooth bacteria got pneumonia and died. The mice injected with the rough bacteria lived. Griffith suspected this had occurred because the smooth coat protected the bacteria from substances produced by a mouse's own defenses against disease,

its immune system. The rough bacteria, he thought, were invaded and killed by these substances of the immune system. This led Griffith to wonder whether something inside a human or an animal could cause the bacteria to change. He tried injecting different combinations of smooth and rough bacteria into mice, hoping to find some pattern that would provide a clue to how the illness could be stopped.

In the course of these experiments, Griffith heated some of the smooth bacteria until he killed them. Then he mixed the dead smooth bacteria with the live but harmless rough bacteria. He injected this mixture into a mouse. To his surprise, the mouse got pneumonia and died. When Griffith examined the dead mouse, he found that it no longer had dead smooth bacteria and live rough bacteria. All the bacteria were live smooth bacteria, just like the bacteria that had killed the mice in his first experiment.

Griffith knew that the dead bacteria could not have come to life again. Rather, he concluded that the live rough bacteria had absorbed something from the dead smooth bacteria. That substance, he thought, had caused the rough bacteria to develop the smooth coat, which protected it from the mouse's immune system.

Griffith did not have any idea what this substance was that had passed from the dead smooth bacteria to the live rough bacteria. He did not try to find out, either. His main interest was in curing pneumonia, and this path of research did not seem to be leading in that direction. So, he abandoned it and moved on to other experiments in an effort to find the cure. Unfortunately, he was killed by German bombers as he worked in his laboratory in London thirteen years later, during World War II. Griffith never found a cure for pneumonia, and he never learned why the rough bacteria had been transformed into smooth bacteria in his experiments with mice.

Finding DNA

But other scientists who had heard about Griffith's work were intrigued. Three American scientists—Oswald Avery, Colin MacLeod, and Maclyn McCarty—decided to continue the bacteria experiments. They wondered if the genes for a smooth coat had somehow passed from the dead smooth bacteria

to the live rough bacteria. If that had occurred, the three scientists realized, they might be able to identify the substance in the cell that had carried out this transfer. They decided to search for this substance even though they knew that identifying it would be a long, difficult process. Each cell contains thousands of different substances. To find out which one carried the gene, they would have to separate and remove all the substances from the cell. Then, they would have to inject each of the substances one at a time into the rough bacteria. If the rough bacteria developed a smooth coat, they would know that the substance they had just injected was the one carrying the genetic information.

Avery, MacLeod, and McCarty began this difficult work in 1933 and continued it for ten years. They knew they had to be extremely careful to purify each substance completely so they would be able to determine what was responsible for carrying the genetic information. After separating each substance, they injected

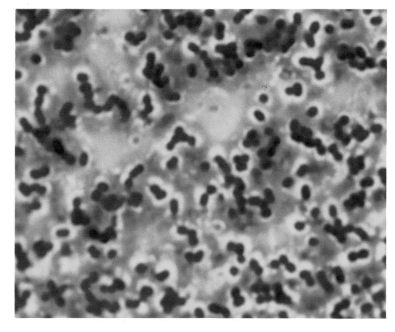

Although he did not know it at the time, British scientist Frederick Griffith observed DNA at work while searching for a cure for pneumonia. His work involved pneumonia-causing bacteria, one type of which is pictured here.

Geneticist Colin MacLeod worked with Avery and McCarty on research that led to the discovery that DNA is the prime carrier of genetic information.

it into the rough bacteria to see if bacteria would develop a smooth coat. One of those substances did cause the bacteria to change. It was a substance called deoxyribonucleic acid, or DNA.

DNA was not new to the scientific world. Chemists had known since Mendel's time that cells contain several kinds of chemical compounds called nucleic acids. They had named one variety that was found only in the nucleus of the cell DNA. A similar substance found in the cytoplasm, the jellylike material outside the nucleus, was called ribonucleic acid, or RNA. Both chemicals are found in the sex cells as well as all other body cells and were known to pass from generation to generation. For this reason, it seemed likely to Avery, MacLeod, and McCarty that DNA or RNA had something to do with heredity.

This suspicion prompted them to take a closer look at the two chemicals. When they did, they found that DNA was a molecule that contained a sugar called deoxyribose, a molecule of phosphorous and oxygen called a phosphate group, and four other molecules called bases. The bases were named adenine, thymine, guanine, and cytosine, usually identified just as A, T, G, and C. RNA had the same ingredients except that a base called uracil, or U, took the place of thymine.

The Key to All Life?

These chemicals were the same in all the living organisms the scientists studied. Humans, trees, cats, bacteria, mice, and pea plants all had the same chemicals in their cells. Many scientists thought that DNA and RNA must be responsible for heredity. Their next question was how the four bases in DNA and RNA could make so many different forms of life. They set out to learn how such a small number of ingredients could ensure that some pea plants had purple flowers and some had white flowers or that humans grew hair and cats grew fur.

Other scientists, though, dismissed the idea that DNA and RNA were responsible for heredity. These scientists thought it more likely that proteins in the cells contained the genetic instructions. They knew that proteins are huge molecules found in all living things, and they knew that proteins take many forms. These scientists were also aware that proteins are made up of twenty building blocks called amino acids. They thought that a molecule with twenty ingredients was more likely to direct life than either DNA or RNA, with their four base ingredients.

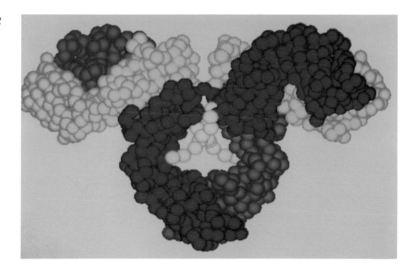

Many scientists thought DNA and RNA were responsible for heredity. But others thought proteins, depicted here by a computer model, were more likely to contain the genetic instructions for life.

Among those who thought DNA and RNA were responsible for heredity were two teams of scientists, one conducting research at Cavendish Laboratory in Cambridge, England, and

Scientist Rosalind Franklin researched the structure of DNA at Oxford University at the same time Watson and Crick were working at Cambridge.

the other at King's College in London. The Cavendish team consisted of James Watson of the United States and Francis Crick of Great Britain. Both men showed brilliance in their work. They had a knack for looking at scattered data from several scientists and pulling this information together to form an overall picture. The King's College team consisted of two British scientists, Rosalind Franklin and Maurice Wilkins. While working in France, Franklin had developed a reputation as an expert in X-ray crystallography, a process that revealed much about the structure of molecules. Wilkins was already using X rays to study the DNA molecule when Franklin arrived in London in 1950.

Looking for Clues

Almost from the beginning, Franklin and Wilkins clashed. Whatever the cause, the dislike they developed for each other made it impossible for them to work together. Many historians have speculated that Franklin and Wilkins might have discovered the structure of DNA if they had worked well together, studying their results and discussing their theories, as Watson and Crick did. Unlike Franklin and Wilkins, Watson and Crick immediately developed a good relationship. Their shared fascination with DNA made them inseparable coworkers. This relationship as well as persistence and inspiration ultimately enabled Watson and Crick to discover the structure of DNA, which helped explain the mechanism of heredity.

All four researchers had studied many other scientists' work, and the bits and pieces of evidence all pointed to DNA. The four were determined to find out exactly how DNA carried genetic information from one generation to the next. But first, they would have to find out what DNA looked like. Each team used different methods to accomplish this task.

Franklin and Wilkins tried to take pictures of DNA using X-ray crystallography. This technique involves passing an X-ray beam through a crystallized molecule. As the X ray passes through the molecule, the beam scatters into a

Oxford genetics researcher Maurice Wilkins worked with Franklin to determine DNA's structure, but personality conflicts prevented their success.

pattern. Each type of molecule creates its own unique pattern. The pattern can then be recorded on photographic film. This technique had already provided

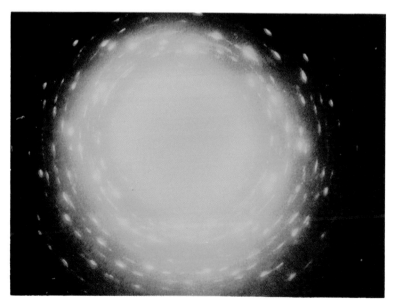

A DNA molecule diffracts an X-ray beam into this characteristic pattern. This technique of photographing molecules, known as X-ray crystallography, was used by Franklin and Wilkins in their attempts to determine the shape of a DNA molecule.

Famed genetics researchers James Watson (left) and Francis Crick pose with a model of a DNA molecule.

clues about the structure of molecules too small to see under a conventional microscope, and Franklin and Wilkins thought it might do the same for DNA.

Piecing together other scientists' results, Watson and Crick tried to build a model of DNA with sticks and rods that looked much like a child's Tinkertoys. Both pairs of scientists knew about the four bases—A, T, G, and C. They had some evidence that the molecule twisted in a spiral, or helix, and they thought it was probably made up of several strands.

A Frantic Race

In his book, *The Double Helix*, Watson describes those years as a frantic race to discover the structure of DNA. Many scientists around the world realized that the structure of DNA would reveal the nature of genes. And the scientist who

showed what a gene was made of would solve one of the biggest mysteries of life. He would be able to describe the invisible mechanism by which traits are passed on from one generation to the next.

Watson and Crick built several models of DNA but later found other bits of evidence that proved their models were wrong. Then, in 1953, Watson went to visit Wilkins at King's College. Without Franklin's knowledge, Wilkins showed Watson an X-ray photograph Franklin had taken of the DNA molecule. Wilkins and Franklin knew the photograph was one of the best ever taken of DNA, but they still did not understand DNA's structure. For Watson, though, seeing the photograph was like having a final piece of the puzzle fall into place. "The instant I saw the picture my mouth fell open and my pulse began to race," Watson recalled in *The Double Helix*. The photograph

RUNGS ON A LADDER: THE STRUCTURE OF DNA

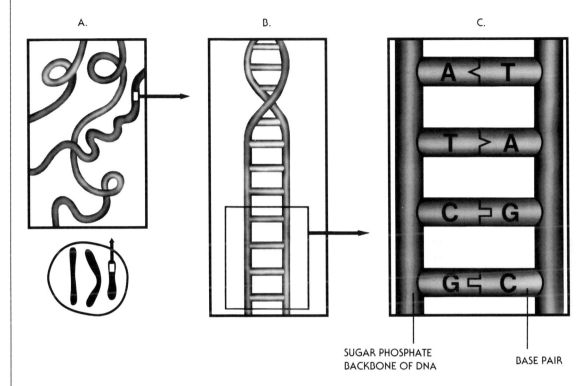

A.   B.   C.

SUGAR PHOSPHATE
BACKBONE OF DNA   BASE PAIR

A. A chromosome is a chain-like strand of DNA, which contains many genes.

B. When the chromosome is greatly magnified under a microscope, it looks like a long ladder that is twisted into a double helix. The twisting allows these amazingly long strands to fit inside a single, tiny cell.

C. The sides of the DNA ladder are made of sugar and phosphate molecules. Between the two sides are rungs made up of the four base pairs—AT, TA, GC, and CG. A single strand of DNA may contain billions of rungs. The different arrangements of these four base pairs are codes that call for different combinations of amino acids. Each group of base pairs that identifies a single amino acid is called a gene. Amino acids combine to make up proteins, which, in turn, combine to form the endless variety of features that make up every living thing.

was a pattern of blurred dots, but it confirmed Watson and Crick's belief that DNA was a helix. The location of the dots caused Watson to realize that the molecule must consist of two chains. Watson rushed back to his laboratory and began to work on a new model for DNA, combining the knowledge that it was a double helix with previous discoveries about its chemical composition. After a few false starts, Watson fitted the pieces together into a tall contraption of sticks and balls. The result, he said, was "very pretty."

NATURE VERSUS NURTURE

Long before the science of genetics developed, people wondered whether human traits and behavior were caused by heredity or by their environment. They could see that some traits, such as hair color and general body shape, were obviously passed on from parents. But with many other characteristics, the relative influence of nature or upbringing was puzzling.

Knowing which characteristics are influenced by environment and which are a fact of birth can help society determine how to improve people's lives.

A 1990 study at the University of Minnesota explored these questions to try to determine how much influence heredity and environment have on behavior. A team of researchers led by Thomas J. Bouchard Jr., studied fifty-two sets of identical twins who had been separated as young children and raised apart. Identical twins have exactly the same genes, so the researchers could easily measure the effect of genes on the twins' development. The fact that these genetically identical twins were raised in different homes enabled the researchers to also measure the effect of environment.

When Bouchard and his team reunited these long-separated pairs of twins, they were often astonished at the similarities between siblings. Several pairs dressed alike, had the same glasses, and wore their hair in the same styles. Members of one pair were both volunteer fire captains; two other siblings were both deputy sheriffs. One pair came to the university with identical items in their shaving kits. The researchers found other pairs of twins in which both people were superstitious, cried easily, or feared water.

Overall, Bouchard's group found that twins tend to have similar mannerisms, gestures, and speed and tempo in talking. They found eleven personality traits, such as shyness or adventurousness, that seem to be inherited. And they found a genetic link to certain attitudes, such as belief in religion or support for a particular political position. Perhaps most important, they found that general intelligence is strongly affected by genetic factors. This study suggests that nature may be a stronger force than previously thought. But more research is sure to follow, as scientists continue to seek answers in this difficult debate.

The DNA Ladder

DNA looks like a ladder that has been twisted into a spiral, Watson and Crick said in the paper they published on April 25, 1953. The upright poles of the ladder are sugar and phosphate molecules. The rungs of the ladder are pairs of the four bases. Each base can form a pair with only one other base: A bonds with T, and G with C. So the rungs on the ladder are always either AT, TA, GC, or CG.

Watson and Crick said that each ladder of DNA contains an immense number of rungs. It was later learned that a strand of human DNA can have up to twelve billion rungs. If a strand of DNA from a human cell could be stretched out, it would be six feet long, thousands of times longer than the cell containing it.

Even though a molecule of DNA has only four base pairs, those four can be combined in an almost limitless number of ways over the length of such an immense chain. The pairs combine the same way the letters in our alphabet combine to form words. The alphabet has only twenty-six letters, but they can be combined in many different ways to express all the concepts the human mind can imagine. The huge number of possible combinations of bases along a strand of DNA convinced many doubting scientists that this simple chemical really could be responsible for so many different forms of life.

Watson and Crick's model made so much sense that many scientists switched from other fields in order to study DNA. Once they knew what DNA looked like, they could figure out how it carried its genetic message from parent to offspring. With that basic knowledge, genetics could be put to use in agriculture, medicine, and many other sciences. It was as if the door to the secret world of heredity had been thrown wide open and scientists just had to walk in and take whatever information they needed.

The triumphant story of Watson and Crick contains one sad note, however. Rosalind Franklin, whose picture of DNA had given Watson the key to the door, died of cancer in 1958 at the age of thirty-seven. She was not even mentioned in 1962, when Watson, Crick, and Wilkins received the Nobel Prize for discovering the structure of DNA.

How DNA Works

During the decade after Watson and Crick's discovery, scientists came to understand how DNA directs life. In general, DNA works with RNA to control the formation of the many different kinds of proteins that make living things so different from one another. Step by step, DNA translates the instructions encoded in the sequence of its four bases into the millions of kinds of proteins that exist in living things.

The DNA in the nucleus begins the process by making what is called messenger RNA. First, the DNA ladder unzips down the middle, separating the matched pairs of bases. On one half of the ladder, the exposed bases link up with bits of RNA that are floating around in the nucleus. The new pairs must follow the same rule of pairing: Adenine bonds with uracil, which is RNA's substitute for thymine, and guanine goes with cytosine. In this way, the RNA bases are arranged in a specific

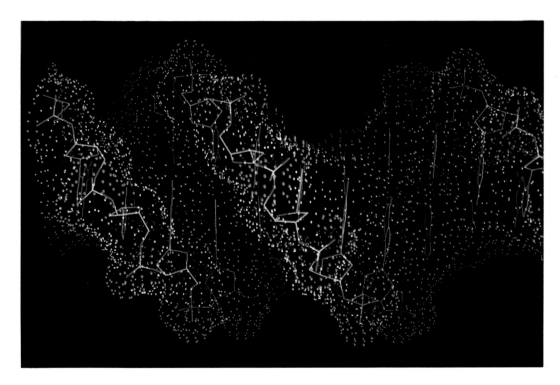

A computer model of DNA. DNA contains the code for the millions of types of proteins that make up all living organisms.

order that mirrors the DNA chain. Once the entire DNA molecule has unzipped and RNA bases have moved into position, the new chain of RNA breaks away. The DNA chain zips closed again and twists back into the double helix.

Next, this newly created strand of messenger RNA moves outside the nucleus and into the cytoplasm. There, it waits for another variety of RNA called transfer RNA to line up with the exposed bases. These bits of transfer RNA are attached to amino acids that they carry behind them. So, when the transfer RNA base hooks up with the messenger RNA base, amino acids are lining up behind. The amino acids form proteins. Each different protein is formed by a different series of amino acids. The different types of proteins are put together according to the spe-

cific arrangement of bases in RNA, which was, in turn, created by the arrangement of bases in DNA. A gene is the section of DNA that produces a single protein.

The construction of proteins is like the construction of a house: DNA is the architect's blueprint; the genes are the drawings for individual features in the house, such as stairs or hallways; RNA is the builder; the amino acids are the building materials; and the proteins are the house.

DNA controls all of life because it contains the code for the millions of types of proteins that make up organisms. The complexity of life is governed by this constant arrangement inside the cell: The four bases of DNA and RNA can pair up only one way: A with T (or U), and G with C.

Genetics and Human Health

When DNA is working well, the human body is like a highly organized factory that runs smoothly in spite of the frenzy of activity within it. But the smallest malfunction in DNA can cause disease, disability, and even death. Malfunctioning genes are known to cause more than four thousand disorders, ranging from a mildly troubling problem such as color blindness to life-threatening diseases that kill children soon after birth.

Scientists now understand that genetic mistakes happen fairly regularly as the cell goes through the complex processes of mitosis, meiosis, and protein production. The code of life is so precise that even a small change in DNA, RNA, or the proteins can have devastating consequences. For example, the chromosomes are supposed to divide during meiosis so that the sex cells have only half the original number of chromosomes. But sometimes the division does not work right, and the sex cell ends up with an extra chromosome. A baby born with an extra chromosome will suffer serious health problems. The most common defect is Down's syndrome, which can cause severe mental retardation and early death.

Another type of mistake can occur when the DNA chain unzips to create messenger RNA. Sometimes, the order of the DNA bases gets scrambled, or an extra base is inserted or deleted. These kinds of changes, called mutations, can occur spontaneously or they can be caused by environmental factors such as excessive exposure to X rays. The cell has special proteins called enzymes that watch for mistakes in the DNA and usually fix them before they cause problems.

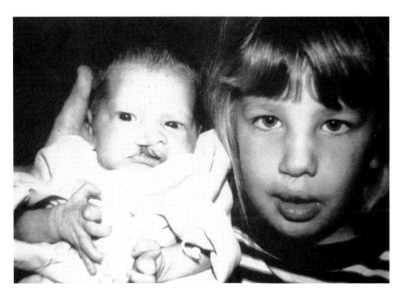

Birth defects, such as cleft palate, sometimes result from genetic mistakes. But unlike cleft palate, many genetic conditions cannot be surgically repaired. So, scientists hope genetic research will lead to ways of preventing or repairing malfunctioning genes.

But sometimes, the changes slip through and become a permanent part of the genetic code. When the code gets scrambled in the DNA, the wrong proteins will be made. Many genetic diseases result from the formation of incorrect proteins. The genes that make these proteins are often passed on through many generations of a particular ethnic or racial group along with genes for other characteristics, including eye shape and hair color. An example is Tay-Sachs disease, which mostly affects Jewish people whose ancestors came from Eastern Europe. Babies born with Tay-Sachs do not have the correct protein needed to digest fats. Without this protein, the fats build up and destroy nerve cells, causing

This child was born with Down's syndrome, a genetic disease that can cause severe mental retardation and early death.

great suffering and death, usually by age five.

Dealing with Disease

The rapid gains in knowledge about genetics during the past fifty years have provided doctors with three different approaches for dealing with genetic diseases. They can counsel couples whose families have a history of genetic disease about their risks of having children with the disease. They can test women who are pregnant to see if the fetus has a genetic disease. And they can, in some cases, provide treatment.

The first approach, genetic counseling, begins when a couple has reason to be concerned about having a child with a genetic disorder. They might be worried because they have already had one child with a disease. Or perhaps their parents, grandparents, or other relatives had a genetic disease. Some people may seek genetic counseling because genetic diseases are common to others with the same heritage. The threat of Tay-Sachs disease, for instance, may cause some Jewish people to seek genetic counseling before having children.

When a couple goes to a doctor for genetic counseling, the doctor will take a family history to determine a pattern of inheritance for the disease. The doctor may also take blood samples from the couple and study the chromosomes in these cells to see if the genetic defect is present. If tests show that the man or woman carries the gene for a certain disease, the doctor can calculate the odds that their children will get it. In this way, the couple can make an informed decision about whether they will have children.

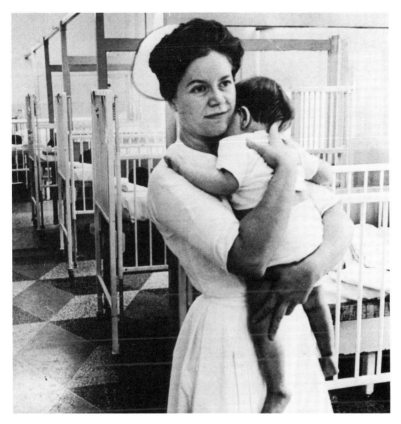

(Left) This child suffers from Tay-Sachs disease, a genetically inherited condition that often kills its victims before they reach the age of five.

(Below) Prospective parents receive genetic counseling from their physician in order to minimize the possibility of producing defective offspring.

The second approach to genetic disease is called prenatal diagnosis. *Prenatal* means "before birth." With this approach, doctors can determine if the fetus a pregnant woman is carrying has a genetic disorder. The most common prenatal test is called amniocentesis, and it is performed when a woman is about four months pregnant. This procedure requires the doctor to insert a long needle through a woman's abdomen and into the uterus, where the fetus is growing. The doctor then draws off some of the fluid, called amniotic fluid, surrounding the fetus. The amniotic fluid is then analyzed for abnormalities in the chromosomes or for chemical clues to genetic diseases. More than fifty disorders can be identified with this method. With the information obtained from prenatal testing,

a couple can decide whether to continue the pregnancy.

The third approach to genetic disease is treatment. The success of this approach varies greatly. Some genetic disorders can be stopped altogether through treatment; others cannot be helped at all.

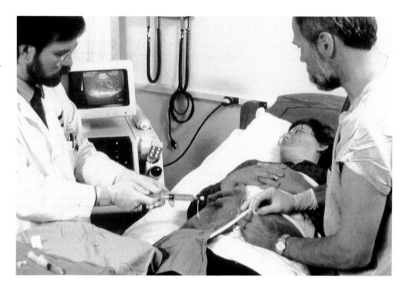

A pregnant woman undergoes amniocentesis, a test that can determine the presence of more than fifty types of congenital diseases in a developing fetus.

One of medicine's greatest success stories in the treatment of genetic diseases has been with the inherited disease phenylketonuria, or PKU. PKU afflicts one in ten thousand babies in the United States. It is caused by a faulty gene that fails to make a chemical the body needs to digest proteins in dairy products, eggs, and fish. Without this natural chemical, these proteins will accumulate in a child's body and prevent the brain from developing properly. The child can become severely mentally retarded.

In the 1960s, scientists learned to detect PKU in newborn babies by checking their blood for the chemical that digests proteins. If the baby lacks the chemical, doctors will immediately put the child on a low-protein diet. By staying on this diet until the brain is fully developed, at about age six, the child is likely to develop normally.

The sickle-shaped red blood cells in this blood sample are caused by defective hemoglobin that causes the normally plate-shaped cell to collapse.

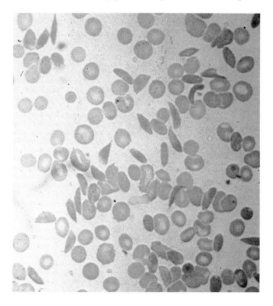

Sickle-Cell Anemia

Unfortunately, many genetic diseases do not have such easy cures. In fact, many cannot be treated at all, even though doctors know exactly what causes them. This is the case with one of the most common and painful genetic disorders, sickle-cell anemia. Scientists first became aware of the disease early in the twentieth century, and much of the work on sickle-cell anemia provided insight into the workings of the genes. Linus Pauling, an American chemist, began to study sickle-cell anemia in 1945. He wanted to find the cause of the disease and suspected that this discovery would provide a clue to the way genes direct life.

Sickle-cell anemia primarily affects black people. In the United States, one in five hundred black infants is born with it. Children with the disease suffer extreme fatigue and weakness, severe pain in their joints, and even brain damage. They usually die before reaching their teens.

Most people have cushion-shaped red blood cells that carry oxygen from the lungs to the tissues. Sickle-cell anemia results when red blood cells collapse into a sickle, or crescent, shape. This occurs when the red blood cells are low in oxygen. When these cells collapse, they get stuck in tiny blood vessels called capillaries, blocking the flow of blood, and they break apart easily. As a result, the blood cannot transport enough oxygen to the tissues to keep them healthy.

Pauling thought that the red blood cells collapsed into sickles because of a defect in the protein hemoglobin, which is the part of the cell that carries oxygen through the bloodstream to the tissues. Pauling studied sickle cells for more than three years to find out if the problem was in the hemoglobin. Finally, he proved that sickle-cell hemoglobin is different from normal hemoglobin. But he was unable to identify the differences.

Six years later, though, the British scientist Vernon Ingram discovered that a molecule of hemoglobin was made up of 560 amino acids. Ingram found that

Biochemist Linus Pauling determined that sickle-cell anemia was caused by defective hemoglobin, the oxygen-carrying protein in red blood cells. Though the cause of sickle-cell anemia is now known, there is still no cure.

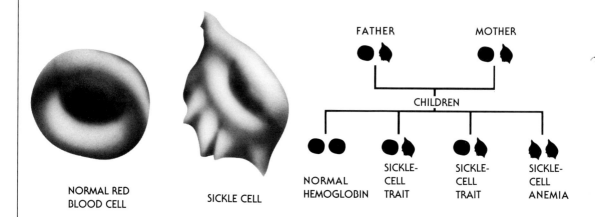

SICKLE-CELL ANEMIA: A GENETICALLY TRANSFERRED DISEASE

NORMAL RED BLOOD CELL

SICKLE CELL

FATHER   MOTHER

CHILDREN

NORMAL HEMOGLOBIN

SICKLE-CELL TRAIT

SICKLE-CELL TRAIT

SICKLE-CELL ANEMIA

Sickle-cell anemia is one of the most deadly of the diseases transferred by human genes, called congenital diseases. Sickle-cell anemia causes the body to produce an abnormal form of hemoglobin, the part of the red blood cell that carries oxygen. Blood cells containing the abnormal hemoglobin often collapse into the shape of a sickle, or crescent. As a result, the victim of the disease becomes gradually weaker, and rarely survives beyond the age of forty.

The gene that causes sickle-cell anemia is recessive, so to get sickle-cell anemia, a child must inherit one gene for the disease from each parent. Persons who inherit the gene from only one parent will not get the disease, but they remain carriers and can pass it on to their children. As the chart shows, if each of two parents carries one normal hemoglobin gene and one sickle-cell gene, their children have one chance in four of inheriting the disease.

Sickle-cell anemia occurs most often among black people of African ancestry, although it also occurs in some Mediterranean and other groups. Researchers have noted that there may be a connection between sickle-cell anemia and malaria, an often fatal tropical disease. The parasite that causes malaria cannot live in the blood of a person with sickle cells. This means that a person who has only one recessive allele for sickle-cell anemia probably would not contract malaria and will not get sickle-cell anemia either.

just one of these amino acids was different in the sickle cells. The other 559 amino acids were the same as those in a normal red blood cell. Scientists were startled to learn that such a tiny flaw inside a cell could kill a human being.

Later, after scientists learned the process by which DNA creates proteins, they found the flaw in the DNA sequence that caused the faulty hemoglobin to be produced. Yet even with that knowledge, doctors cannot prevent or cure the disease. The most they can do is calculate the odds of a couple having a child with sickle-cell anemia. Calculating the odds is relatively easy. Doctors know that sickle-cell anemia is a simple recessive trait. It is inherited in the same way white flowers were inherited in Mendel's peas. This means that people can carry the recessive gene for the disease without suffering the disease themselves, just as the first generation of Mendel's peas carried the recessive white allele without showing it. But when two people who carry the recessive sickle-cell gene have a child together, they run the risk that the two recessive alleles will combine in the baby, which will then develop the disease.

Cancer and the Oncogene

Although scientists have learned a great deal about genetic disorders in the past fifty years, they are still perplexed by many diseases. One of the most puzzling is cancer, the dangerous and uncontrollable growth of cells. The American Cancer Society has estimated that a child born in the United States in 1987 has a one-in-three chance of developing cancer during his or her lifetime and a one-in-five chance of dying from it. Cancer afflicts people of all ages and ethnic backgrounds. Researchers now think that many types of cancer are hereditary, but the relationship between genes and cancer is not entirely clear. It appears that many genes are involved in the development of cancer and that environmental factors, such as diet and radiation, also play a role in determining who gets cancer and who does not.

Many scientists think that some types of cancer are caused by special genes called oncogenes. These oncogenes appear to be a harmless part of the cell under normal conditions. In fact, some oncogenes may be responsible for starting the cell growth necessary for a fertilized egg to develop into a baby. After this growth is completed, some scientists think, the oncogenes switch off but remain in the cells. Some research has shown that the oncogenes may be accidentally switched on again when irritated by environmental factors,

Many scientists think some types of cancer are caused by genes called oncogenes. A microscopic view of cancer cells is shown here.

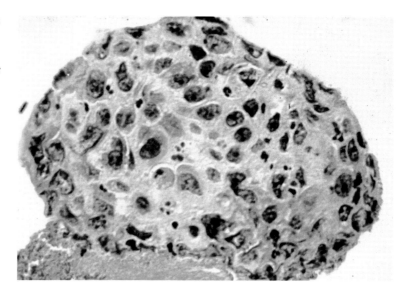

The relationship between genes and cancer still is unclear. Researchers hope to learn more in the coming years. Pictured here, a view of a cancer cell.

resulting in uncontrolled cell division that causes tumors to grow throughout the body. Medical researchers hope to find a way to switch the oncogenes off and, in this way, stop the spread of cancer.

The search for the cause of cancer shows that science still has a long way to go in the quest to understand the link between genes and disease. But the answers are within reach, and developments within the past few years have given scientists even better weapons to use in the search for causes and cures. The next era in genetic research will involve the use of technology to lessen human suffering. Through a series of recently developed techniques, scientists can already create medicines never before available. They can even sometimes replace defective genes that are causing diseases. And someday, they hope to be able to identify every gene that is responsible for a disease and then go into the cells to repair those that are defective.

Genetic Engineering

Once scientists understood the relationship between genetics and disease, they began to search for ways to prevent or cure diseases by altering the genes responsible for disease. This goal required many years of basic research into the complex rules of genetics. As these rules became better understood in the 1970s, scientists developed new techniques that allowed them to change the characteristics of an organism by changing its genes. They learned how to cut strands of DNA and how to remove and insert genes from the DNA of one organism into the DNA of another. These techniques are called genetic engineering. The use of genetic engineering to develop commercial products is called biotechnology.

Rapid Progress

Genetic engineering has progressed rapidly in the past twenty years. Scientists in many disciplines have found ways to use these techniques. The results of their work are increasingly evident today and will be even more important to society in the future. This is particularly true in the fields of medicine and agriculture.

Genetic engineering had its beginnings in 1970, when scientist Hamilton Smith and his coworkers at Johns Hopkins University in Baltimore, Maryland, found a reliable way to cut DNA. Because DNA is so tiny, scientists had

Almost every day genetic engineers create and manufacture new products in laboratories like this one.

been unable to accurately cut it with scissors or knives. So, Smith and his colleagues explored other ideas. During their experiments, they discovered that an enzyme found in one strain of influenza bacteria could be used to cut DNA. When the researchers inspected the pieces of DNA cut by this enzyme, they noticed something interesting. The cuts always occurred after the same sequence of bases. Every time this enzyme saw the bases GTTAAC, for example, it would snip that sequence from the strand of DNA. Through further experiments, they found that this same enzyme could recognize the GTTAAC sequence in any type of organism, from bacteria to humans. No matter which organism the researchers used, the enzyme always cut the DNA in exactly the same place, after the sequence GTTAAC.

These discoveries prompted an intensive search for other enzymes that would cut DNA. Today, scientists know of about one hundred enzymes capable of cutting DNA. The enzymes used in this technique are called restriction enzymes, and their discovery brought science one step closer to the goal of replacing or deleting faulty genes.

Stanford University geneticist Stanley Cohen helped pioneer gene splicing in the 1970s.

Manipulating DNA

In 1972, four scientists decided to try these new techniques to move DNA from one organism to another. Stanley Cohen and Annie Chang at Stanford University in Palo Alto, California, and Herbert Boyer and Robert Helling at the University of California, San Francisco, were not trying to create a useful product or cure a disease. They just wanted to see if it were possible to transfer genes, and figure out the best way to do it. But they also knew that if they succeeded, their work would lay the foundation for practical applications in the future.

The four scientists began looking for clues in the cells themselves. They knew that some bacteria cells contain an extra chromosome called a plasmid, which is a small and relatively uncomplicated ring of DNA. They also knew that the plasmid copies itself and sends a copy to each new cell created during cell division. The four scientists realized that plasmids would be a good vehicle for transporting new genes into cells. They devised a technique to do this and then set the experiment up using a toad.

First, they used restriction enzymes to cut a gene from a toad's DNA. Then, they used another restriction enzyme to cut open the circular plasmid of the bacterium. Finally, using a third type of enzyme, they attached the piece of toad DNA to the broken ring of the bacterium's plasmid and joined the ends to close the ring again. This procedure is called gene splicing because a piece of DNA is combined with, or spliced into, another strand of DNA in the same way the strands of one rope can be woven into the strands of another. The altered DNA produced in this procedure is called recombinant DNA

RECOMBINING DNA

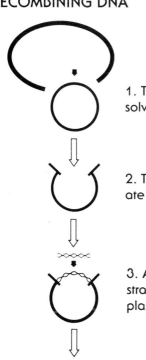

1. The cell wall of a bacterium is dissolved, and a plasmid spills out.

2. The plasmid is cut, using the appropriate enzymes.

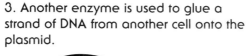

3. Another enzyme is used to glue a strand of DNA from another cell onto the plasmid.

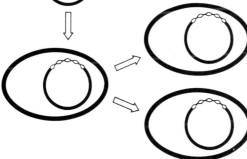

4. The plasmid is reabsorbed by a bacteria cell, which divides normally, producing numerous clones with the altered genetic characteristics.

Geneticists use special kinds of proteins, called enzymes, to cut genes from one DNA strand and attach them to another strand. Through experimentation they have discovered about one hundred of these special enzymes called restriction enzymes. Each one cuts or attaches DNA strands only before or after a certain sequence of bases. Using this knowledge, geneticists can remove a segment of genes from the chromosome of one cell and attach it to a DNA strand in a different cell.

They have been particularly successful producing clones, or copies, of altered bacteria cells. These cells contain a strand of DNA called a plasmid, which is easily removed from a bacteria cell and later absorbed by another bacteria cell. Since bacteria reproduce very rapidly, the cells with the altered plasmids will quickly produce thousands of copies of themselves, each with the altered plasmid and genetic traits that it carries.

because the DNA of two organisms has been recombined.

When the group's genetically altered bacterium divided, the plasmid carried the toad gene into the new cells. These bacteria divided further, providing many identical copies of the original altered bacterium. This process of producing identical copies of a living organism is called cloning.

These discoveries rocked the scientific world. Not only had the four scientists combined DNA from two different organisms but they also had done it with two different species, a toad and a bacterium. In nature, a toad and a bacterium would never mate, and their genetic material would never mingle. These experiments broke through what had been previously thought to be an impenetrable barrier between species. This led scientists the world over to realize that entirely new forms of life could be created in a laboratory.

The possibilities seemed endless: Recombinant DNA might be used to create superproductive crop plants, bigger livestock, and better medicines. These new techniques could even be used to replace or repair the defective genes that caused illness in humans. In every field of biology, and even in some of the other sciences, genetic engineering seemed to have fantastic potential uses. Scientists immediately began trying to apply these new procedures to their own areas of expertise.

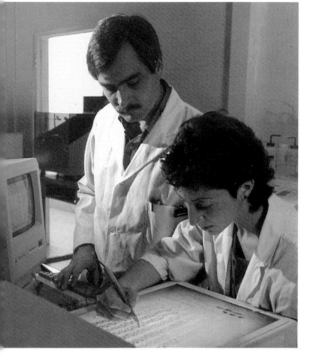

(Above) Scientists study recombinant DNA in a modern laboratory. Experiments conducted in the 1970s opened new avenues for recombinant DNA technology. (Right) A genetics researcher examines a bottle containing a human cell culture as part of a research experiment with recombinant DNA.

This laboratory worker is harvesting human insulin produced by genetically engineered bacteria.

Intense Competition

Throughout the 1970s, an intense competition raged among scientists to produce a human gene. Many wanted to be first, not only for the recognition they would surely receive but also because they saw the great commercial possibilities in products that could help people. Attention focused on the gene that provides the code for the production of a protein called insulin, which helps convert sugar to energy in the pancreas. People with a disease called diabetes do not produce enough insulin on their own, so they have to give themselves injections of insulin taken from pigs. This pig insulin causes negative side effects in many people who take it. Scientists thought they could help many diabetics by producing human insulin through genetic engineering.

Herbert Boyer again played an important role in this development. Boyer, working with Arthur Riggs and Keiichi Itakura, began by analyzing human insulin and determining its amino acid makeup. Next, the Boyer group determined the sequence of bases in the gene that produced insulin. Once they had identified the gene, they were able to duplicate it in the laboratory. They took this synthetic gene and spliced it into the DNA of a bacterium. Finally, they cloned the altered bacterium to produce many identical copies.

The scientists found that the insulin gene produced insulin in its new bacterial hosts. This meant that the altered bacteria could be produced in vast quantities to provide large amounts of human insulin. A new biotechnology firm, Genentech, began growing the altered bacteria in huge vats and harvesting the insulin. In December 1980, the newly manufactured insulin was tested in a

An electron micrograph shows the structure of a human insulin molecule produced by recombinant DNA methods.

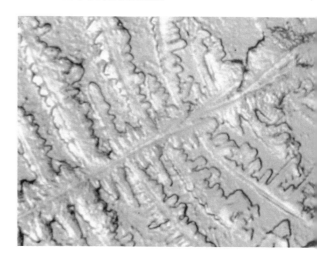

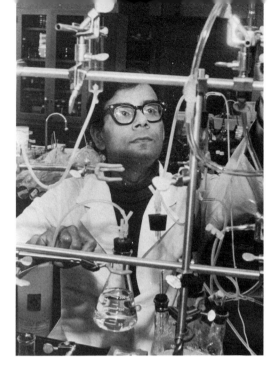

Ananda Chakrabarty works in his laboratory. The U.S. Supreme Court ruled in 1980 that Chakrabarty could patent a genetically engineered bacterium.

A Genetically Engineered Mouse

In 1984, the science of genetics took another huge step forward when Philip Leder and Timothy Stewart, scientists at Harvard University in Cambridge, Massachusetts, genetically engineered a mouse. This marked the first time scientists had succeeded in changing the genetic makeup of a higher organism, and it represented a step toward the ultimate goal of altering human genes.

Leder and Stewart's immediate goal was to use the genetically engineered mouse in cancer research. The two men decided to splice an oncogene suspected of causing cancer into the mouse. They reasoned that if the oncogene were solely responsible for cancer, all the altered mice would get cancer at the same time and with the same severity. But if the mice did not get cancer uniformly, the scientists would know that factors other than the oncogene were to blame.

Harvard University geneticist Philip Leder genetically engineered a mouse to be used in cancer research.

human subject. This injection marked the first time a genetically engineered product was used in medicine, and this type of insulin is still used today.

Another milestone in genetic engineering was passed in 1980 when the U.S. Supreme Court ruled that Ananda M. Chakrabarty, a scientist at General Electric, could patent a genetically engineered bacterium. Chakrabarty had developed a strain of bacteria that could digest oil more quickly than natural bacteria. He hoped they could be used to clean up oil spills. The Supreme Court ruling on Chakrabarty's bacterium had a profound effect on genetic research. The ruling meant that a company with a genetically engineered organism could receive a patent that would give the company exclusive rights to sell the organism. Companies saw that they could make money in genetic engineering, and many increased their research efforts.

ONCOMOUSE: THE WORLD'S FIRST GENETICALLY ENGINEERED MOUSE

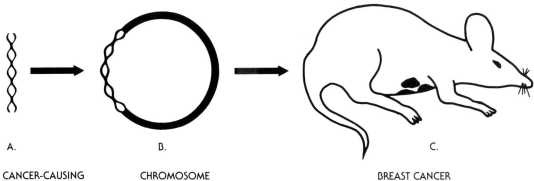

A.

CANCER-CAUSING
ONCOGENE

B.

CHROMOSOME
FROM MOUSE EGG

C.

BREAST CANCER

In 1984, Harvard University scientists Philip Leder and Timothy Stewart produced the world's first genetically engineered mouse. They named it Oncomouse because they extracted oncogenes from cancerous cells known to cause breast cancer (A) and inserted these oncogenes into the DNA of fertilized mouse eggs (B). Shortly after they were born, a high percentage of the Oncomice began to develop symptoms of breast cancer (C). This made the genetically engineered mouse particularly useful for cancer research.

In the first experiment, they determined the DNA sequence of a gene associated with breast cancer. They constructed several pieces of DNA with this oncogene sequence. Then, they inserted the pieces of constructed DNA into the DNA of fertilized mouse eggs. The genetically altered mice were born and began to develop cancer in their breast tissues, at a much higher rate than mice without the gene. However, the tumors were not all the same. This suggested that oncogenes were responsible for cancer but that other factors also played a role.

Harvard sought to patent the mouse, which was called Oncomouse, so that no one else could genetically engineer mice with cancer genes. In April 1988, the patent office granted Harvard exclusive rights to the Oncomouse. Many kinds of new technology receive patents, but this was the first time a living, breathing animal had been patented.

Dr. Philip Leder explains his genetically engineered Oncomouse.

Bubble Babies

During the 1980s, scientists conducted research involving the many diseases caused by defective or missing genes. They hoped to develop a procedure for inserting a working gene into a sick patient to replace the gene causing the illness. This type of treatment is called gene therapy.

Dr. W. French Anderson and Dr. R. Michael Blaese at the National Institutes of Health decided to try gene therapy on children whose white blood cells lack an enzyme called ADA. White blood cells help prevent infections by engulfing foreign matter, such as bacteria, that enters the body. A deficiency of ADA prevents the immune system from working, and even the slightest sickness can kill a child. As a result, these children must live in sterile tents, or bubbles, to protect them from germs. Their disease, ADA deficiency, is often called "bubble baby syndrome." A few cases have been successfully treated by transplants of bone

National Institutes of Health researcher R. Michael Blaese pioneered gene therapy for immune system deficiencies.

marrow, where the white blood cells are produced. But it is extremely difficult to find an appropriate match for bone marrow transplants. Most children with ADA deficiency die of infection by age five.

Six-year-old Jared Reisman had such severe allergies he had to live in a plastic air bubble or wear a protective, air-filtering suit. Doctors in Denver, Colorado, tried to identify the genetic defect that had caused Jared's problem.

Anderson and Blaese began their gene therapy by identifying the gene that produces ADA in normal people. Then, they cut the sequence of DNA from a normal human cell. Next, they removed some white blood cells from a child with the disease. They spliced the gene for ADA into the child's white blood cells. The altered white blood cells began producing ADA. The researchers sought permission from the federal agencies that oversee genetic research to begin gene therapy on a four-year-old girl with the disease, and in 1990, permission was granted. Anderson and Blaese gave the child a transfusion of the altered white blood cells. Initial reports, three months after the injections, said that the therapy was working. The child's cells were producing their own ADA. Most white blood cells live only a few weeks to a few months under any conditions, so the researchers know that the altered cells will have to be replaced frequently. The final judgment about the effectiveness of gene therapy rests with time. If the child has a normal life span, then the therapy may be declared a success.

Increasing Food Production

During the same time these advances were occurring in medical science, agricultural scientists were busy trying to use genetic engineering to increase food production. One company developed genetically altered bacteria to protect crops from frost damage. Some years, frost has caused more than one billion dollars' worth of damage to food crops. Many plants have naturally occurring bacteria that enhance the

Oranges and other fruits and vegetables are susceptible to damage from frost. This damage can run into billions of dollars.

formation of ice crystals, and these ice crystals can destroy many kinds of vegetables and fruits. Scientists identified the gene in the bacteria that controlled their ability to form ice and then removed it from the bacteria. The altered bacteria were called "ice-minus" bacteria. In 1987, the ice-minus bacteria were sprayed on fields of strawberries in California. The strawberries were protected from frost damage at much colder temperatures than usual.

A technician from a genetic engineering company sprays strawberries with genetically engineered bacteria that help the berries resist frost damage.

Genetic engineers have also found ways to make livestock more productive. At the St. Louis, Missouri-based Monsanto Company, one of the largest researchers in agricultural genetic engineering, scientists were able to increase milk production in cows by 10 to 25 percent. They did this by identifying a chemical that cows naturally produce in small amounts called bovine growth hormone. They ana-lyzed the amino acid makeup of bovine growth hormone and then determined the sequence of DNA that produced the hormone. Once that was identified, they were able to cut that particular piece of DNA out of the cow cell and splice it into a bacterium. The bacterium began to produce bovine growth hormone. So, the bacterium was cloned and then grown in huge quantities. Finally the hormone was

The cotton boll on the right is genetically engineered to resist the devastating assaults of the boll weevil that damaged the boll on the left.

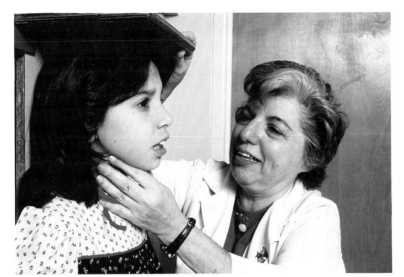

A health care worker monitors the progress of a girl being treated with a recombinant DNA growth hormone. For some, genetic engineering offers the promise of a healthy, normal life.

extracted from the bacteria. Monsanto officials hope the hormone will win federal approval for use by commercial dairies by 1992.

These discoveries in medicine and agriculture are just a few examples of the many advances made in genetic engineering within the past few years. In addition, scientists have created bacteria to recycle trash into fuel, hormones to cure dwarfism in children, a strain of soil bacteria to kill a type of worm that attacks the roots of corn plants, and a tobacco plant that will not be hurt by weed killers. Many other products of genetic engineering are still being tested in the laboratory. Within a few more decades, genetic engineering may be so commonplace that people may take for granted these forms of life never found in nature.

The Ethics of Genetics

Developments in genetics in the past twenty years have given scientists immense new powers. Today, through the techniques of genetic engineering, scientists are able to create forms of life that do not exist in nature. They are able to bypass some of life's most basic rules, making changes overnight that would take nature generations to accomplish.

Surveys show that most Americans welcome these advances and consider them to be beneficial to society. But everyone familiar with genetic research, including the researchers themselves, acknowledge that the science poses many ethical questions that must be considered.

Wise Use or Abuse?

The most basic question is whether genetic engineering itself is an ethical use of genetics research. Some people worry that human beings do not understand enough about life to risk tampering with it. They fear that society does not realize what an awesome power genetic engineering gives to individuals, and they worry that not everyone will use this power wisely. The most outspoken opponent of genetic engineering in the 1980s and 1990s has been Jeremy Rifkin, head of an organization in Washington, D.C., called the Foundation on Economic Trends. Rifkin's group has repeatedly filed lawsuits to try to stop genetic engineering experiments. He has

Genetic engineering critic Jeremy Rifkin believes many genetic engineering experiments are unethical.

also urged Congress to pass laws requiring stricter regulation of genetic engineering. One of Rifkin's chief concerns is that genetic engineering gives power over life to corporations that are interested only in making a profit. He worries that genes will become commodities to be bought and sold. Rifkin has warned that living things could become "engineered products with no greater intrinsic [essential] value than microwave ovens." Rifkin prefers to think that life is sacred and should not be deliberately changed to suit someone's notion of what would make a better organism. Genetic engineering, Rifkin warns, is as powerful and as dangerous as the atomic bomb.

Advocates of genetic engineering agree that it is a powerful tool. But they argue that society does have the wisdom to use this power well. They cite the many good results that have already come from genetically engineered products. For instance, insulin is available for diabetics who get sick from animal insulin. Bacteria can clean up oil spills. Plants can be protected from late frosts that might otherwise wipe out a crop. These early genetic engineering experiments, some say, prove that scientists are motivated by a genuine desire to help humanity and improve society.

Scientists themselves argue that genetic engineering is not likely to be misused because most work is subject to review by other scientists. This review occurs either when a scientist first applies for a grant to conduct research or when that person publishes the results of the experiments in a scientific journal. If other scientists think the work is unethical, the researcher will be forced to make changes in the project. Most scientists believe that this system, called peer review, helps protect against abuses of genetic knowledge.

Safety Concerns

One of the most important examples of peer review in genetic research occurred when recombinant DNA techniques were first discovered. In 1970, Stanford University scientist Paul Berg was planning an experiment that involved putting the gene for a swine virus called SV40 into the bacterium *Escherichia coli*. This bacterium, abbreviated *E. coli*, normally lives in the intestines of humans. By infecting the human *E. coli* with the virus, Berg hoped to learn more about how viruses turn normal cells into cancerous cells. Several other scientists who heard about Berg's plans became alarmed. They feared that *E. coli* containing the virus might accidentally escape from Berg's laboratory and infect humans. They feared that the altered bacteria could cause cancer in people.

As Berg listened to these concerns, he realized that the new techniques of genetic engineering could pose hazards to people. Berg agreed to head a committee that would study these biological hazards and make recommendations for keeping experiments safe. In July 1974, Berg's committee of scientists wrote an open letter to all researchers asking them to stop their work until

In 1974, Stanford University biochemist Paul Berg formed a committee to study safety issues related to genetic engineering experiments.

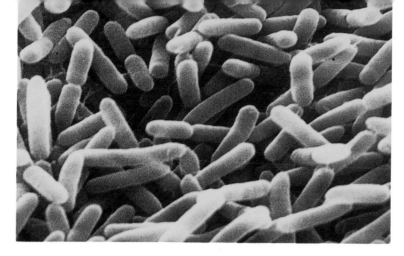

Escherichia coli *bacteria. E. coli is a common strain of bacteria found in the human intestinal tract. Some scientists are concerned about genetically altered* E. coli *escaping from laboratories and infecting the public.*

safety standards could be created by the National Institutes of Health, the federal agency that oversees much health-related research. In February 1975, 150 scientists from around the world met at the Asilomar Conference Center in California for a four-day discussion of genetic engineering. The scientists concluded that research should be conducted only in laboratories equipped with special facilities that would prevent the escape of genetically engineered organisms. The scientists also recommended that a strain of *E. coli* should be genetically altered so that it would not be able to survive outside the laboratory and

that this new strain should be the only type used in genetic research experiments. In this way, scientists could be doubly safe. Hazardous bacteria would not be likely to escape, and if they did, they would die quickly. Sixteen months after the Asilomar conference, the National Institutes of Health also proposed the same safety standards.

These rules calmed fears about genetic hazards for several years. But in 1983, Steven Lindow, a scientist at the University of California, Berkeley, proposed taking a strain of genetically altered bacteria out of the laboratory for testing in the environment. Lindow had

University of California scientist Steven Lindow (right) wanted to test a genetically altered bacteria outside the laboratory. A lawsuit seeking to halt the test failed.

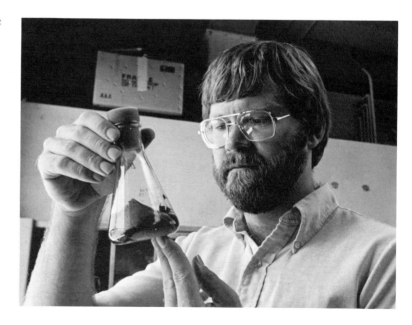

discovered a way to protect crops from frost damage. He had discovered a gene in bacteria that normally live on plants that cause ice to form. He removed the gene from the bacteria. He planned to spray these altered bacteria, called ice-minus bacteria, on a plot of potatoes in northern California. The National Institutes of Health approved the project, but then Jeremy Rifkin intervened. Rifkin's organization filed a lawsuit to prevent the outdoor experiment. Rifkin argued that the ice-minus bacteria would spread to wild plants and disrupt their natural cycle. He also predicted that if the bacteria spread into the atmosphere it could prevent rain from forming in clouds because water vapor needs a tiny ice crystal before it will condense into raindrops. The lawsuit stalled Lindow's experiment until 1987. Finally, the courts approved the outdoor tests, and the ice-minus bacteria were sprayed on crops. The bacteria did prevent frost damage, and no bad side effects to the environment were observed.

The controversy over the ice-minus bacteria made the government extremely cautious about approving genetic engineering experiments outside the laboratory. Concern for the environment has prevented many genetic engineering research projects from being completed.

Past Misuses

Even greater concerns about genetics experiments exist in the area of human research because the world knows too well the abuses that have occurred in the past when scientists tried to improve human life. In the 1920s, a movement to improve human beings, called eugenics, became popular in Europe and the

Nazi leader Adolf Hitler had more than six million persons killed in an attempt to purge the German people of genetic traits he considered inferior.

United States. Many people believed that they could create a better society if they could prevent people with bad genetic traits from having children. Supporters of the eugenics movement proposed that sterilization surgery be used to prevent certain people from reproducing. In the United States, thirty states passed laws requiring sterilization of people who were mentally ill, alcoholic, or afflicted with a disease called epilepsy. Under these laws, about twenty thousand people were sterilized against their will. Today, we know that alcoholism, epilepsy, and many kinds of mental illness can be treated or cured.

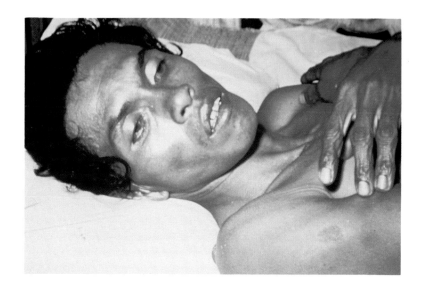

Malaria kills many people in Africa. But researchers have found that Africans with one recessive gene for sickle-cell anemia are immune to malaria. Can genetic engineering eliminate one deadly disease without increasing the risk of another?

In fact, many people with these ailments have made valuable contributions to society.

While the eugenics movement was growing in the United States, a similar sterilization program began in Germany in the 1930s under that country's new leader, Adolf Hitler. In one year, fifty-six thousand people were sterilized. But Hitler did not stop there. Next, he began a program to create what he considered a superior race of tall, blue-eyed blonds. Obsessed with the idea that dark-haired people would pollute the genes of his "superior" race, Hitler ordered the killing of Jews, Gypsies, and other groups of people who did not fit his idea of a superior race. Before he was defeated in World War II, Hitler killed about six million people because of their genetic makeup.

The horrors of Nazi Germany made people realize the danger of trying to genetically improve the human race. Most people today would agree that no one should have the power to decide which traits should be weeded out of society and which traits should be promoted as superior.

Difficult Decisions

Yet such decisions are still made every day on a small scale because of advances in genetics. When a pregnant woman learns through prenatal testing that the fetus she is carrying has a genetic disorder, she must decide whether to give birth to the child or have an abortion. Essentially, this is a decision about whether the genetic disorder is a trait that should be eliminated from her family to prevent the suffering of the child and family. Many people oppose abortion in such cases. They argue that children with genetic disorders such as Down's syndrome or cystic fibrosis bring love and joy into the world despite their disabilities.

Other people oppose abortion in such cases because they think genetic disorders may play an important role in the larger scheme of populations. Because genetic disorders evolve over many lifetimes, the effects are not apparent. As evidence of this position, they point to sickle-cell anemia, the genetic defect that causes red blood cells to collapse and break apart, preventing oxygen from getting to the tissues and resulting in major damage to the organs. The disease is common among people of African ancestry. Most children who inherit two recessive alleles for sickle cells will get the disease and usually die young.

Scientists have recently learned, however, that people in Africa who carry only one recessive allele for the disease may actually be better off than people who do not carry the trait at all. That is because the parasite that causes malaria, an often fatal tropical disease, cannot live in the blood of a person with sickle cells. So people with one allele for sickle-cell anemia do not get malaria. And because they have only one recessive allele, they do not get sickle-cell anemia, either. In other words, this genetic defect that is devastating for some people is actually beneficial to other people. This suggests to some scientists that genetic diseases may serve a function that no one really understands. And, they say, people should not disturb this larger pattern of life by tinkering with human genes.

On the other side of this debate are people who know the anguish caused by an incurable genetic disease. These people do not want to bring a child into the world only to have the child suffer a painful, early death. They seek genetic counseling, prenatal testing, and abortion in the case of "genetic disorder" as a compassionate way to prevent tragedy.

New Procedures Raise New Questions

Another ethical question concerns a procedure for families with a history of genetic disease. In this procedure, a doctor can remove eggs from the mother's ovaries and fertilize them in a test tube with the father's sperm. The fertilized eggs divide. When they have eight cells, one cell from each embryo is removed, and the DNA in these cells is analyzed for genetic defects. An embryo that does not have any genetic defects will be implanted in the mother, and she will become pregnant and deliver the baby normally. Embryos that do have genes for the genetic disease will be thrown away. The first people to use this procedure were at risk of having children with Tay-Sachs disease, a painful disease that usually kills a

A human embryo develops according to innate genetic programming. Some people believe that genetically altering embryos would improve people's chances for good health later in life.

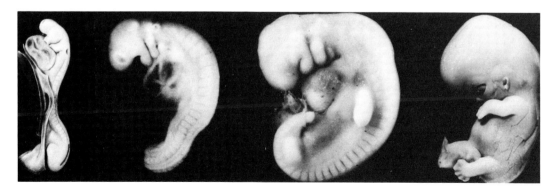

child by age five. Other genetic diseases may be prevented with this procedure as soon as scientist can identify the specific genes responsible for each disease.

This new procedure has raised questions about how it might be used in the future, as scientists identify genes for characteristics other than disease. If the genes for eye color, height, and intelligence could be identified, for example, couples could check an embryo to see if it had the traits they considered desirable. If it did not, they could throw the embryo out and start again. Or they could have a dozen embryos analyzed and choose the one with the genes they liked best.

These possibilities disturb many people who think that nature should be allowed to dictate the genetic traits of children. Even some of the scientists who developed the procedure worry that it could be abused if people someday use it to determine trivial matters such as eye color. Some experts, however, believe that families should be able to manipulate their genes with whatever technologies are available. They see genetic testing, for whatever reasons, as a personal decision.

Another problem area in genetics is the ability to screen people to determine if their genes make them susceptible to certain problems. Some people, for instance, have a gene that makes them likely to get lung disease if they are exposed to a certain industrial chemical. Genetic screening could provide this kind of information. Such information could be helpful since people usually do not know they have this genetic trait unless they are exposed to the chemical. With this information, the person susceptible to lung disease could just stay away from jobs in which expo-

sure might occur. But suppose the only job this person can find requires use of the chemical? Some people would argue that such a person should be free to take the job, even though it might someday cause lung disease. Others would argue that such a person should be prevented from taking the job for health reasons, no matter what the difficulty of finding another job.

Remembering the Past, Determining the Future

These are difficult questions that society will have to answer soon. Some companies already use genetic screening to determine whom to hire for jobs, and many more companies are thinking about starting genetic screening. To some corporations, genetic screening makes good sense because it can help prevent the hiring of sick employees, which costs the company money. Sick employees use more health-care benefits and are absent from work more often. But many doctors and lawmakers fear that genetic screening could lead to discrimination in hiring, promotion, and health care. In the worst case, they say, mandatory genetic screening could create a "genetic underclass" of people who would be forced into certain careers and prevented from achieving the equal opportunities that are the foundations of American society.

This has already happened in the United States. In the 1970s, the Air Force Academy tested black applicants for the recessive sickle-cell trait, which does not affect the person carrying it. But the Air Force argued that even one recessive allele might cause the person to have weak red blood cells. These

In 1981, an Air Force cadet sued the Air Force Academy for denying blacks with the sickle-cell gene admission to the academy. Academy officials had feared that the black pilots' red blood cells would collapse at high altitudes.

weak red blood cells might collapse when the oxygen supply becomes extremely low, as it often does in high-flying Air Force jets. Because of this fear, which was not scientifically proven, the academy denied admission to black people who carried the recessive allele. A cadet sued the U.S. Air Force in 1981, saying the policy discriminated against blacks. The academy stopped genetic testing and again began to accept black people with the sickle-cell allele.

Many of the questions society has already faced about genetics will become even more pressing in the future, as scientists learn more about genes. The potential for great abuse to individuals, ethnic groups, society in general, and the natural environment cannot be ignored. At a meeting in Valencia, Spain, in October 1988, two hundred scientists from twenty-four countries agreed to "acknowledge their responsibility to help insure that genetic information will be used only to enhance the dignity of the individual." It is up to the rest of the world to make sure scientists keep that promise.

CHAPTER 8

Fighting Crime with Genetics

In the late 1980s, police began to use genetics in the fight against crime. They learned that every human cell contains a person's full package of DNA and that no two people, except identical twins, have the same DNA. This unique genetic pattern is called a DNA fingerprint.

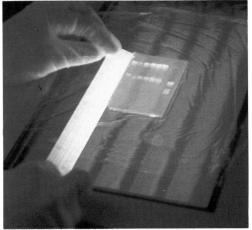

(Above) A police scientist searches DNA patterns, hoping to find a match with a suspected criminal's DNA. (Right) Police officers at the scene of a crime need only find hair, skin, or blood, from which a lab can extract a DNA fingerprint.

DNA fingerprints can be used to identify criminals in many types of crimes, including murders, rapes, and thefts. When a murder occurs, for example, police search for blood, hair, skin, or other tissue at the scene of the crime or take it from a suspect's possessions. They look for these tiny bits of evidence because each strand of hair, each piece of skin, or each drop of blood is made up of thousands of cells containing a person's unique pattern of DNA. Even a few cells can be used to connect a person to a crime.

After collecting this cellular evidence, police send it to a laboratory that specializes in DNA analysis. If police find a strand of hair, for example, lab technicians extract a chain of DNA from the cells of the hair strand. Next, they use restriction enzymes to cut the

If these scientists determine that a crime suspect's DNA pattern matches that of tissue samples found at a crime scene, the suspect's guilt is virtually assured because of the slim chance that someone else would have the same DNA pattern.

DNA into segments at specific points in the chain. Because everyone's DNA is different, the cuts will occur at different points, and these segments of DNA will be different lengths for each person. The segments are then lined up according to size and photographed. The picture looks like black stripes on a white background, similar to the bar codes found on packaged goods at the grocery store.

The lab will also cut and photograph pieces of DNA from a hair strand, blood sample, or other tissue taken directly from the suspect. A suspect's pattern of stripes can then be compared with the pattern of stripes obtained from the evidence. If the patterns look the same, the laboratory can then calculate the likelihood that the suspect committed the crime. They do this by using data for the number of times the suspect's genetic patterns occur in his or her racial group.

Because racial and ethnic groups share certain genetic patterns, this helps to narrow the population. Most of the time, the odds against the suspect are immense. For example, the laboratory might say that there was one chance in one billion that someone besides the suspect could have the same DNA pattern. Such huge odds have usually been enough to convince juries that the suspect was guilty.

Just one drop of blood contains millions of cells all harboring a unique DNA pattern.

Establishing Guilt and Innocence

DNA fingerprinting was developed in Great Britain in 1985 and first used in the United States in 1987. Since then, it has played a role in solving many serious crimes. In the United States, it has been used in more than four hundred cases in thirty-eight states. At least eighty people have been convicted of crimes because of DNA evidence.

This new crime-solving tool is most helpful in cases where police cannot find other evidence, such as weapons or actual fingerprints. In Ventura, California, a woman named Lynda Patricia Axell was accused of stabbing to death a sixty-three-year-old maintenance man named George White. White was killed in 1988 when he tried to stop a robbery at the Top Hat Burger Palace where he worked. No one saw the murder, and Axell denied doing it. But police found sixty long, brown hairs at the scene. They sent the hairs to a laboratory for

Through DNA fingerprinting, a single shaft of hair (magnified here) can provide strong evidence linking a suspect to a crime.

DNA fingerprinting. Then, they took some long, brown hairs from the suspect and sent those to the same laboratory. The pattern of stripes from the DNA in the suspect's hair matched the pattern of stripes from DNA in the hairs found at the scene of the crime. The laboratory estimated that there was only a one-in-six-billion chance that someone else could have the same DNA pattern. Axell was convicted of murder and attempted robbery and sentenced to twenty-five years to life in prison.

DNA fingerprinting can even be used to solve crimes in which the victim cannot be found. Police can link a suspect to the missing person by collecting DNA samples in the hair, blood, or other tissue from the parents of the missing person. This is because every person receives half of his or her DNA from each parent. Once scientists know the parents' DNA patterns, they can establish a biological relationship between parent and child.

This occurred in a Johnson County, Kansas, case in which a painting contractor named Richard Grissom Jr., was accused of killing three women. Two of the women, Christine A. Rusch and Theresa J. Brown, lived in an apartment building where Grissom worked. The third woman, Joan Marie Butler, lived in a nearby city. The women had disappeared, and police suspected Grissom had murdered them in June 1989 and hidden their bodies. Police found the suspect driving a car that belonged to Butler. When they inspected the car, they found a few drops of blood in the trunk. Police scraped up the blood cells and sent them to a laboratory for DNA fingerprinting. Next, they got blood samples from Butler's parents, Ralph and Jada Butler, and sent those samples to the laboratory. The

DNA fingerprinting helped convict Richard Grissom of murder. Drops of blood provided an important clue in the case against Grissom.

parents' DNA contained patterns that matched the patterns from the blood in the car. Experts testified at the trial that the blood must have belonged to a child of these parents. Although a lot of evidence tied the painter to the three missing women, the drops of blood were the strongest evidence that violence was involved in their disappearance. Grissom was convicted of murdering the three women and was sentenced to life in prison.

DNA fingerprints can be used to prove innocence, too. Sometimes, police will take a DNA sample from a suspect and compare it with the DNA in blood, hair, or tissue that they found at the scene of a crime. When the DNA patterns do not match, police assume they have arrested the wrong person.

In London, police reported that DNA tests proved the innocence of suspects in 28 percent of the cases where DNA fingerprinting was done. In New York, DNA fingerprints kept an insurance company office worker out of prison by proving that he had not committed a crime. The man was accused of raping the daughter of a city police officer. The woman had picked the man out of a lineup and identified him as the attacker. But his DNA fingerprints did not match the DNA fingerprints of tissue found at the scene of the crime. Police immediately freed the wrongly accused man.

Reliable Tool or Faulty Technique?

Despite the many cases where DNA fingerprints have been used, the technique is still in dispute. Some lawyers, judges, and scientists have criticized the use of DNA fingerprints in court for several reasons. First, questions have been raised about the procedures used in the DNA fingerprinting laboratories. If technicians are not extremely precise in conducting the tests, mistakes can occur. For example, contamination of the evidence by bacteria or dirt could make prints from different people look the same, scientists say.

In addition, some scientists question the calculations used by the laboratories to show the probability that someone else might have the same genetic pattern as the suspect. The laboratories calculate probability by finding the frequency of certain genetic markers within the suspect's racial group. These frequency figures are obtained from blood samples, such as those taken at a

(Right) A scientist compares images of DNA. Identifying and matching DNA patterns is called DNA fingerprinting. Skin cells, magnified below under a microscope, contain the unique genetic information that can help identify a specific individual.

had the wrong suspect. But another expert in the same case calculated the odds and found that 1 in 78 people could have the same DNA pattern. These odds show that the chances of another person having the same DNA pattern are much greater, and so the suspect's guilt becomes less certain.

Another concern about DNA fingerprinting is that there could be an innocent explanation for a match between a suspect's DNA and evidence from a crime scene. The suspect might have been at the scene earlier and not have committed the crime, for example. Or the criminal who knows about DNA fingerprinting could leave someone else's hair at the scene of the crime to shift suspicion toward that other person.

Many lawyers also think that DNA fingerprinting is just too complicated to explain to juries. They argue that scientists still dispute the reliability of the technique and that jurors should not be expected to determine which conflicting claim is accurate. Until scientists agree that genetic testing is foolproof, some lawyers say, it should not be used in trials.

blood bank. Some scientists say the number of blood samples is too small to accurately determine the frequency of these genetic markers within the much larger racial group. As a result, the laboratory may overstate the odds against error in a match. In one New York case, for example, a DNA testing firm estimated that only 1 in 100 million people could have the same DNA pattern as the suspect. This would seem to indicate very little chance that the police

Debate Continues

For the first two years that DNA evidence was used in the United States, judges did not question it. Then, in a 1989 murder trial in New York, two prominent scientists told the judge that the laboratory's procedures were not strict enough to guarantee accuracy. They complained that laboratories doing DNA fingerprinting were not regulated or inspected by anyone and that this lack of supervision could lead to abuses. The judge agreed and refused to permit the use of DNA fingerprints in the trial.

Since then, DNA fingerprints have been challenged in many cases. Appeals courts in eight states have approved the use of genetic evidence in trials. But in three other states, courts have prohibited the use of DNA fingerprints as evidence. The issue is pending in many other states, including California, where it has been challenged by the woman who was convicted on the basis of DNA fingerprints taken from the long, brown hairs found at the scene of the crime.

One of the most important legal decisions so far was made by a federal judge in Ohio who conducted lengthy hearings on the reliability of DNA fingerprints. After listening to the testimony of many experts, the judge concluded that DNA fingerprints can identify criminals or prove innocence. He found that a large number of scientists accept the technique as accurate and concluded that it should be permitted as evidence in criminal cases. An important report on the subject of genetic fingerprinting came from the Congressional Office of Technology Assessment in 1990. That agency surveyed hundreds of scientists and officials and reported that most experts agreed that the basic principles underlying DNA fingerprinting are valid and can be used in criminal cases.

Debate on this new technology is likely to continue. Eventually, the courts or lawmakers may step in and decide rules for using DNA fingerprints. But until they do, police and prosecutors will continue to use genetic evidence when they have no other way of convicting a criminal.

The Future of Genetics

Recent discoveries in genetics have been impressive, but the future promises to deliver even greater advances. The knowledge gained during the past century has just begun to be applied to human problems. Many more practical uses of genetics research, particularly in medicine, agriculture, and industry, are likely to be developed.

Mapping Human Genes

The most ambitious plan for the future of genetics is a massive, federally funded effort called the Human Genome Project. Although scientists know that every human being has forty-six chromosomes, they do not know exactly how many genes are in each chromosome. Scientists estimate that the human genome, or collection of chromosomes in which almost all human genes reside, consists of from 50,000 to 100,000 genes. The goal of the Human Genome Project is to identify every one of those genes and then draw a map of their location on the chromosomes. This process will be tedious. Even with many scientists all over the country working on the project, it will not be completed until the year 2003. And it will cost U.S. taxpayers a total of three billion dollars.

But supporters of the project say the results will be well worth the money, time, and effort. James D. Watson, one of the discoverers of the structure of DNA, is the director of the Human Genome Project. He predicts that creating a map of human genes will enable doctors in the future to diagnose, prevent, and treat genetic disorders because doctors will know exactly which genes are responsible for those disorders.

Before the project began, the locations of about 500 genes were known. By late 1990, 5,150 genes had been identified. New discoveries are pouring in nearly every week.

The human genome map ultimately will identify all human genes and pinpoint their locations on the chromosomes. Below is a segment of the human genome map.

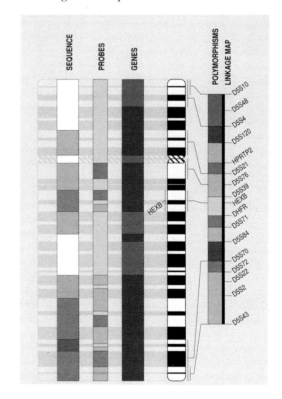

A scientist uses an electronic video monitor to analyze a set of chromosomes. Doctors hope someday to be able to scan the chromosomes of a fetus and repair any defective genes before the child is born.

The first targets of the investigation are the genes that cause diseases. Scientists already know the genes responsible for many terrible illnesses, such as cystic fibrosis and muscular dystrophy. They expect they may find genetic causes of depression, alcoholism, cancer, and many other ailments that plague humanity.

In time, they may also find the location of the genes that determine physical characteristics, such as hair color or the shape of a nose. Eventually, they may even be able to identify genes that determine personalities, religious convictions, and other attitudes.

The gene map will help scientists with another goal for the future—finding a way to fix defective genes. They envision a day when they will be able to scan the chromosomes of a fetus, find any defective genes, and repair them before the baby is even born. The baby would then develop normally. It could be possible to avoid altogether the suffering caused by genetic defects.

A researcher at the National Institutes of Health interprets a readout of a DNA sequence. Scientists are slowly learning to read the full message contained in such a sequence.

On the Horizon and Beyond

Another innovative experiment that may pay off in the future concerns AIDS, the acquired immunodeficiency syndrome. Although AIDS is not a genetic disease, genetic research may help scientists find a cure for this deadly disease. Dr. J. Michael McCune, who treated AIDS patients at San Francisco General Hospital, became frustrated when he was not able to test possible AIDS cures on animals because the AIDS virus infects only human cells. So, in 1987, McCune decided he would try to produce human immune responses in mice so that drugs could then be tested in these mice.

Dr. J. Michael McCune genetically engineered laboratory mice to adapt them to test AIDS drugs.

McCune did this with the help of genetic engineering technology. He took thymus, liver, and lymph nodes from human tissue and implanted a piece of each organ under the kidneys of young mice. He found that within days, the mouse's blood vessels moved into the human tissue, encouraging it to grow. Within a short time, the tissue developed immune responses much like those occurring in human organs. Today, McCune can infect these humanized mice with the AIDS virus and test various drugs for a cure. He and other researchers are also using this process in efforts to cure other diseases.

In agriculture, genetic advances are expected to come quickly. Researchers have already succeeded in transferring genes from one type of plant to another to create combinations that would never occur in nature. For example, a "pomato" that grows potatoes below ground and tomatoes above has already been produced in the laboratory. Similar oddities may someday be available to farmers and gardeners.

But even more likely are combinations of plant genes that would solve real problems for farmers. For example, most grain plants, such as wheat, require a great deal of nitrogen to grow well. So farmers apply fertilizers containing nitrogen to these crops every year. But legumes, a family of plants that include soybeans, have the ability to make their own fertilizer by taking nitrogen from the air. A gene found in legumes makes this possible. Scientists are trying to take this gene from legumes and insert it into the DNA of grain plants. The result would be a grain that makes its own fertilizer, which would save farmers money on chemical fertilizers.

Legumes grow by making their own fertilizer from nitrogen in the air. Scientists hope to find the gene in legumes that performs this feat and transfer it to grains like corn and wheat. The grains could then make their own fertilizer, too, and farmers would save millions of dollars in fertilizer costs.

Other agricultural problems are being approached in a similar fashion. Some scientists are trying to genetically engineer plants to resist common insects and diseases that destroy millions of dollars' worth of crops each year. Other research focuses on improving the nutritional value of plants used to feed livestock. The genes of corn, a plant high in carbohydrates, might be fused with the cells of soybeans, high in protein, to create a balanced meal in one plant.

Similar changes will be made in livestock, too. Genetic engineering has already produced a hormone to make cows produce up to 25 percent more milk. Agricultural companies are trying to find hormones that will make pigs and cattle grow bigger so they have more meat. Chickens also might be made to produce more eggs with the help of genetically engineered hormones.

Genetics also holds great potential for industry. Scientists are conducting genetic research that might alter or create many types of microorganisms for use in industrial processes. For example, one of the biggest problems facing industry and the environment today is the disposal of toxic wastes. Toxic wastes are hazardous substances that can cause cancer and other illnesses in humans and animals. Many types of toxic waste contain manufactured chemicals that do not occur in nature. Because they are not natural, nature has not provided any organisms to help recycle or break down these chemicals. As a result, these substances can remain dangerous for generations. Proper disposal of these substances costs a great deal of money. And any kind of disposal still presents the possibility that the chemicals will escape and contaminate the environment.

Ananda M. Chakrabarty, the scientist who won the first U.S. patent for genetically engineered bacteria, is working on the problem of toxic wastes. He hopes to find a gene that he can splice into the

Toxic waste poses a threat to the environment and human health. Bacteria genetically engineered to digest such chemicals could help solve this dangerous problem.

DNA of bacteria so that they will be able to break down toxic wastes. The bacteria could then be cloned to provide an unlimited supply of toxic-waste digesters.

These examples of the potential uses of genetics are just glimpses of what is already on the horizon. Beyond that, the science of genetics may cause changes in society that are completely unexpected now. Knowledge of the mechanism of heredity may give humans powers that we can hardly imagine.

Glossary

allele: An alternative version of a gene.

amino acids: The twenty fundamental building blocks of proteins that are linked together to form these proteins according to the instructions contained in the cell's DNA.

asexual reproduction: The reproduction of individual cells or organisms that does not involve fertilization.

bacteria: Simple, single-celled organisms that have no distinct cell nuclei and reproduce asexually.

base: A building block of DNA. There are four types of bases in DNA: adenine, cytosine, guanine, and thymine, usually identified by the first letters of their names.

cells: The building blocks of all living organisms. Each cell contains all the genetic information of an organism.

chromosomes: The long, squiggly strands of DNA found in the cell's nucleus that contain the organism's genes.

cloning: Producing cells and organisms with identical genes.

cytoplasm: The jellylike substance in the cell outside the nucleus.

DNA, or deoxyribonucleic acid: The substance from which chromosomes inside the cell nucleus are formed. Sequences of DNA are genes, which order construction of the millions of types of proteins that make up sequences of all forms of life.

enzyme: A protein molecule that causes a specific chemical reaction in living things.

Escherichia coli: Bacteria found in the human intestines.

eugenics: The attempt to improve human beings by genetic means.

evolution: The change in the genetic makeup of a population over time.

fertilization: The joining of male and female sex cells to form a new life.

gene: A sequence of bases in DNA that directs construction of a protein.

genetic engineering: Deliberately altering or transferring genes from one organism to another.

hemoglobin: The oxygen-carrying protein in red blood cells.

Human Genome Project: A plan for identifying and mapping every gene on the human chromosomes.

hybrid: Offspring produced from the mating of two different varieties of plants or animals.

hybridization: The process of creating a new type of organism by mating two different types of organisms.

insulin: A hormone produced in the pancreas needed for digestion.

meiosis: The division of a cell to form sex cells, in which each daughter cell has half the number of chromosomes of the parent cell.

messenger RNA: A molecule that carries genetic instructions from the DNA in the nucleus to the cytoplasm, where it helps produce proteins.

mitosis: Cell division in which the number of chromosomes is the same in the daughter cells as in the parent cell.

mutation: A change in DNA that can be either spontaneous or caused by environmental changes such as radiation.

natural selection: The survival of the organisms that are best suited to their environment.

nucleus: The central structure in a cell that contains the DNA.

protein: A large molecule made up of amino acids.

RNA, or ribonucleic acid: A molecule similar to DNA but consisting of a single strand of bases in which thymine is replaced by uracil.

sex cell: A sperm (or pollen) or egg cell. The male and female cells come together during fertilization.

sex chromosome: A chromosome that determines an individual's sex.

species: A group of individual organisms that can mate and produce fertile offspring.

transfer RNA: Small molecule of RNA that carries amino acids to the messenger RNA, lining them up according to the DNA instructions, to produce proteins.

For Further Reading

Caroline Arnold, *Genetics: From Mendel to Gene Splicing*. New York: Franklin Watts, 1986.

Isaac Asimov, *How Did We Find Out About the Beginning of Life?* New York: Walker, 1982.

Discovering Genetics. New York: Stonehenge in association with the American Museum of Natural History, 1982.

Aaron E. Klein, *Threads of Life: Genetics from Aristotle to DNA.* Garden City, NY: The National History Press, 1955.

Alvin Silverstein and Virginia Silverstein, *The Code of Life.* New York: Atheneum, 1972.

Works Consulted

Alex Bristow, *The Sex Life of Plants*. New York: Holt, Rinehart & Winston, 1978.

Craig Canine, "Keeping the Promise of Biotechnology," *Harrowsmith Country Life*. September/October 1990.

Elof Axel Carlson, *The Gene: A Critical History*. Philadelphia: W. B. Saunders, 1966.

L. L. Cavalli-Sforza, *Elements of Human Genetics*. Menlo Park, CA: W. A. Benjamin, 1977.

Carol Deppe, "The Joys of Home Hybridizing," *Horticulture*. August 1990.

L. C. Dunn, *A Short History of Genetics*. New York: McGraw Hill, 1965.

Bob Gibbons, *How Flowers Work: A Guide to Plant Biology*. New York: Blandford Press, 1984.

Mahlon Hoagland, *Discovery: The Search for DNA's Secrets*. Boston: Houghton Mifflin, 1981.

Margaret O. Hyde and Lawrence E. Hyde, *Cloning and the New Genetics*. Hillside, NJ: Enslow Publishers, 1984.

Horace Freeland Judson, *The Eighth Day of Creation: The Makers of the Revolution in Biology*. New York: Simon & Schuster, 1978.

Marc Lappe, *Broken Code*. San Francisco: Sierra Club Books, 1984.

Steven Lehrer, *Explorers of the Body*. Garden City, NY: Doubleday, 1979.

Charles Panati, *The Book of Breakthroughs: Astonishing Advances in Your Lifetime in Medicine, Science and Technology*. Boston: Houghton Mifflin, 1980.

Clare Mead Rosen, "The Eerie World of Reunited Twins," *Discover*. September 1987.

Norman V. Rothwell, *Human Genetics*. Englewood Cliffs, NJ: Prentice-Hall, 1977.

Rebecca Rupp, *Blue Corn and Square Tomatoes*. Pownal, VT: Storey Communications, 1987.

Curt Stern and Eva Sherwood, *The Origin of Genetics: A Mendel Sourcebook*. San Francisco: W. H. Freeman, 1966.

David Suzuki and Peter Knudtson, *Genethics: The Clash Between the New Genetics and Human Values*. Cambridge, MA: Harvard University Press, 1989.

James D. Watson, *The Double Helix*. New York: Atheneum, 1968.

Index

abortion
  to eliminate genetic
    disorders, 72-73
adenine, 41, 44, 45, 48
AIDS (acquired immune
    deficiency syndrome),
    84
Air Force Academy
  genetic screening of
    candidates, 74-75
alleles, 24, 25, 34
amino acids, 41, 45, 47
amniocentesis, 51
Anderson, W. French
  gene therapy research,
    64-65
animals
  genetically engineered,
    62-63, 66, 85
  medical testing on, 39,
    58, 60, 62-63, 84
Aristotle
  genetic theory of, 10,
    18
Avery, Oswald
  study of DNA, 11, 40-41
Axell, Lynda Patricia, 78,
    81

bacteria
  genetically engineered,
    58-60, 61, 62, 65, 70-
    71
Bateson, William
  study of genetics, 33-34
Berg, Paul
  committee on safety in

genetic engineer-
    ing, 69
biotechnology, 57
Blaese, R. Michael
  gene therapy research,
    64-65
Bouchard, Thomas J., Jr.,
    46
Boyer, Herbert
  genetic engineering of
    insulin, 61
  gene transfer experi-
    ments, 11, 58, 60
Brown, Theresa J., 78
bubble baby syndrome,
    64-65
Butler, Joan Marie, 78

Camerarius, Rudolf
    Jakob, 10
  plant reproduction
    theories, 15-16
cancer, 55-56
cells
  discovery of, 28
  division of, 28-31
Chakrabarty, Ananda M.
  genetic engineering
    patent, 62, 85-86
Chang, Annie
  gene transfer experi-
    ments, 58, 60
chromosomes
  affected by environ-
    ment, 37-38
  discovery of, 10
  in humans, 30, 34-35

male and female, 34-35
  meiosis and, 30-31
  mitosis and, 29-30
  study of fruit fly, 35-37
cloning, 60
Cohen, Stanley
  gene transfer experi-
    ments, 11, 58, 60
congenital diseases. See
    genetically transferred
    diseases
corn
  improvements in, 13, 85
Correns, Carl, 33
Crick, Francis
  discovery of DNA struc-
    ture, 11, 42-45, 47
crime detection
  using DNA, 76-81
cross-pollination, 15, 16,
    17, 27
cystic fibrosis, 72, 83
cytosine, 41, 44, 45, 48

Darwin, Charles
  evolution theory, 20-21,
    37
  Origin of Species, The, 10,
    20
De Vries, Hugo
  studies on heredity, 32-
    33
deoxyribonucleic acid. See
    DNA
diseases
  caused by DNA defect,
    49-50, 54, 55

caused by environment, 49, 55
DNA (deoxyribonucleic acid)
altering of. *See* genetic engineering
as cause of disease, 49-50, 54
composition of, 41
description of, 47-48
discovery of, 11, 42-44, 47
fingerprinting, 76-77
reliability of, 79-81
use in criminal cases, 78-79
recombinant, 58-60
structure of, 45, 47
studies on, 40-44, 47
Double Helix, The, 44
Down's syndrome, 49, 72

environment
effect on chromosomes, 37-38
effect on human behavior, 46
eugenics, 71-72
evolution, 20-21, 37

farming
history of, 12-13
food production, 13-14
Foundation on Economic Trends, 68
Franklin, Rosalind
work on DNA, 42-44, 47

gametes, 28, 31
Genentech company, 61
genes
definition of, 45, 48
discovery of, 10, 24
on chromosomes, 35-38
dominant, 24, 26, 32, 36
medical therapy and, 11, 61-62, 64-65
mutations of, 37-38
recessive, 24, 26, 32, 36
splicing, 58
transfer experiments, 11, 58-60
genetically transferred diseases
cancer, 55-56
cystic fibrosis, 72, 83
Down's syndrome, 49, 72
muscular dystrophy, 83
phenylketonuria (PKU), 52
sickle-cell anemia, 52-55, 72-73, 74-75
Tay-Sachs, 50, 73-74
treatment of, 52-56
genetic counseling, 50, 55
genetic engineering
agricultural uses, 65, 67
commercial uses, 62, 67, 69, 85
dangers of, 68-72
definition of, 57
ethical concerns, 71-75
medical uses, 61-64, 67, 82-83
of animals, 62-63, 66-67, 85
of insulin, 61-62
patents on, 11, 62-63
safeguards against misuse, 69, 70, 71, 75
genetic screening
for employment, 74-75
genetic testing, 50-52, 72
genome mapping project, 82
Grew, Nehemiah, 10
plant reproduction theories, 14-15
Griffith, Frederick
experiments with bacteria, 39-40
Grissom, Richard, Jr., 78-79
guanine, 41, 44, 45, 48

Harvey, William
genetic theory of, 18-19
Helling, Robert
gene transfer experiments, 58, 60
hemoglobin, 53, 54
heredity
De Vries experiments and, 32-33
evolution and, 21
Mendel's experiments and, 22-27
particulate theory, 18
study of identical twins, 46

Hitler, Adolf, 72
Hooke, Robert, 28
Human Genome Project, 82
hybridization
    process of, 16
ice-minus bacteria, 65, 71
Ingram, Vernon, 53
insulin
    genetically engineered, 11, 61-62
Iraq, 12
Itakura, Keiichi, 61

Leder, Philip
    genetically engineered mouse, 62-63
Lindow, Steven
    development of ice-minus bacteria, 70-71

MacLeod, Colin
    study of DNA, 40-41
Malpighi, Marcello, 14
McCarty, Maclyn
    study of DNA, 40-41
McCune, J. Michael, 84
meiosis, 31, 35
    as cause of disease, 49
Mendel, Gregor, 10
    gene theories of, 24-26
        rediscovered, 33-34
    pea plant experiments, 21-27
microscopes
    used in study of cells, 28-29

mitosis, 29
    as cause of disease, 49
Monsanto company, 66, 67
Morgan, Thomas Hunt, 10
    study of chromosomes, 36-37
Muller, Hermann J.
    study of gene mutations, 37-38
muscular dystrophy, 83

Nageli, Carl, 27
National Institutes of Health
    on genetic engineering, 70, 71
natural selection theory, 21, 37
    see also evolution

Office of Technology Assessment
    DNA fingerprinting survey, 81
oncogenes, 55-56
Oncomouse, 62-63
Origin of Species, The, 10, 20

Pauling, Linus, 53
phenylketonuria (PKU), 52
pistil, 14, 15
plants, 14, 15
    breeding of, 10, 16-17

domestication of, 12
genetically engineered, 65, 67, 84-85
ice-minus bacteria and, 65, 71
improvements in, 13, 16
pea plants
    breeding experiments and, 22-27
pollination of, 15, 16, 17
selection process, 13, 16
proteins, 41, 47

radiation, 37
    as cause of cell damage, 37
restriction enzymes
    cutting DNA, 58, 76-77
ribonucleic acid. See RNA
rice, 13
Rifkin, Jeremy, 68, 71
Riggs, Arthur, 61
RNA (ribonucleic acid), 41, 42, 47
Rusch, Christine A., 78

sickle-cell anemia, 52-55, 72-73, 74-75
Smith, Hamilton, 11
    genetic engineering and, 57
species, 17
Stewart, Timothy
    genetically engineered mouse, 62-63
Tay-Sachs disease, 50, 73-74

thymine, 41, 44, 45, 48
tomatoes
   improvements in, 16-17,
    84
twin studies, 46

uracil, 41, 44, 45, 48
U.S. Supreme Court, 11,
   62

von Tschermak, Erich, 33

Watson, James
   discovery of DNA struc-
    ture, 11, 42-45, 47
   Human Genome
    Project, 82
wheat
   improvements in, 13,
    84-85
Wilkins, Maurice
   work on DNA, 42-44, 47

X rays
   as cause of cell damage,
    37, 49
X-ray crystallography, 42,
   44

About the Author

$\blacksquare$ $\blacksquare$

Lynn Byczynski graduated from the University of
Kansas in 1979 with a B.S. degree in science
journalism. She worked as a newspaper reporter
for ten years. She now teaches journalism at the
University of Kansas and is a free-lance writer. She
and her husband, Dan, spend their summers
growing vegetables, herbs, and flowers for market.
They have one son.

Picture Credits

■■■

Cover Photo by P. Degginger/H. Armstrong Roberts

AP/Wide World Photos, 21, 41, 43 (top), 51 (top), 62 (top), 63 (bottom), 64 (bottom), 65, 66 (top), 73, 75, 76 (bottom)

R. F. Ashley/Visuals Unlimited, 26

The Bettmann Archive, 12, 13 (top), 14 (bottom), 18 (middle), 23, 24, 32 (top)

Biological Images, 13 (bottom), 14 (top), 17 (bottom), 28 (right), 30 (top right and bottom), 49, 55, 56

Bill Bransen/National Cancer Institute, 64 (top)

Cabisco/Visuals Unlimited, 37

R. Calentine/Visuals Unlimited, 35

Christine L. Case/Visuals Unlimited, 85

D. Cavagnaro/Visuals Unlimited, 17 (top)

Centers for Disease Control, 72

Cold Spring Harbor Laboratory, 20 (top), 22 (top), 36, 38, 42 (bottom), 44 (top), 53

M. Coleman/Visuals Unlimited, 50

John D. Cunningham/Visuals Unlimited, 22 (bottom)

Bruce C. Cushing/Visuals Unlimited, 33

R. Feldman, D. McCoy/Rainbow, 42 (top)

Foundation for Economic Trends, 68

Michael Gabridge/Visuals Unlimited, 76 (top), 77 (top)

Kevin Hall, 15, 25, 29, 31, 34, 45, 54, 59, 63 (top)

Historical Pictures Services, 18 (bottom), 28 (left)

Fred Hossler/Visuals Unlimited, 78

Johnson Co. Police Department, 79

Larry Keenan/Genentech, Inc., 61 (top)

R. Langridge, D. McCoy/Rainbow, 48

Library of Congress, 16

Hank Morgan/Rainbow, 60 (top)

K.G. Murti/Visuals Unlimited, 30 (top left)

National Institutes of Health, 52 (bottom), 77 (bottom), 83 (bottom)

News Office of the Harvard Medical Area, 62 (bottom)

Glenn Oliver/Visuals Unlimited, 32 (bottom)

Terry O'Neill, 18 (top)

David M. Phillips/Visuals Unlimited, 19, 70 (top), 80 (bottom)

Science Magazine, 82

Stanford University School of Medicine, 55, 69

A.M. Steigelman/Visuals Unlimited, 40

UC Berkeley, 70 (bottom)

UPI/Bettmann, 46, 71, 86

Visuals Unlimited, 43 (bottom), 51 (bottom), 52 (top), 57, 60 (bottom), 61 (bottom), 66 (bottom), 67, 80 (top), 83 (top)